SHOPPING BAGS & WRAPPING PAPER

GRAPHIC-SHA

Shopping Bags & Wrapping Paper

ISBN4-7661-0482-X

Printed in Japan by Toppan Printing Co., Ltd.

First Edition June 1988

Graphic-sha Publishing Company Ltd.
1-9-12 Kudan-kita Chiyoda-ku Tokyo 102 Japan
Phone 03-263-4310
Fax 03-263-5297
Telex J29877 Graphic

ショッピングバッグ・コレクション

グラフィック社

はじめに

その片田舎の高校に，初めてショッピングバッグを持ち込んだのは，黒い制服のズボンを細く仕立て直してはいてくる一人のハンサムな同級生だった。彼は，学校で決められた黒い手下げ鞄と一緒に，どこかの洋服店の名前が入った白っぽい手下げのペーパーバッグに体操服を入れて学校に下げてきた。それはたちまちクラス中に伝染し，そして学校中に広まった。そのショッピングバッグは，彼の家族のうちの誰かが都会から持ち帰ったと思われるハイカラな雰囲気をもつものだったが，それよりも，只の紙袋に過ぎないものが，学校に持ち込まれることによって異様に新鮮な物体に生まれ変わって見えたことが驚きであった。この一人の田舎の高校生の規則違反のオシャレに似たことは，おそらく当時の日本のあちこちで見られた情景に違いない。今から数十年前，日本が高度経済成長期にさしかかった頃のことである。
マーチャンダイジングに大変革が起こり，やがてスーパーマーケット形式の大量販売形態が全国各地で行なわれるようになる。パッケージングの従来的な形態からの脱皮は，この商品経済の変化と密接な関係をもって劇的に進行していったが，一人の高校生のことも，もちろんそうした経済のダイナミックなうねりに垣間見えた１シーンだったろう。

ショッピングバッグは，その，いわばパッケージングの新しい形態としての簡便さと機能性が，現代の社会に受け入れられた第一の理由であろうが，同時に，企業や店が顧客に対するサービスツールとして使用するという本来の使命から，持ち帰るときに楽しい気分になるような，美しくて，洗練されたものであることが要求される。デザインが重視され，そのことがまた新鮮なインパクトとして社会から注目されるという図式がある。一方，ショッピングバッグは，“歩く宣伝媒体”と呼ばれるように，企業や店のマークやロゴを入れて消費者とのコミュニケーションの役割を果たすが，このときも，デザインが主役の位置に立つ。こうした，ショッピングバッグのヴィジュアル性が強調されるようになったのは，欧米でも15〜20年ほど前にさかのぼるに過ぎないようだが，そのころ東京の六本木や原宿界隈で，今までにない感覚のショッピングバッグが，行き交う人の眼をひくようになる。それは，当時，比較的ひんぱんに海外と往来できる環境にいた人たちや，流行に敏感なファッションモデルたちが，外国から持ち込んで下げて歩いたもので，ニューヨークのシャレたブティックのショッピングバッグであったり，あるいはロンドンのサイケデリックなカラフルなショッピングバッグであったりした。その，街ですれちがいざまに眼に触れた新鮮な光景が，ファッションやグラフィックデザインに関係した人たちの心をとらえる。
しかしながら日本では，実際上，それほど急速には，ショッピングバッグのデザインはホットな関心事とはならなかった。現在でも，ショッピングバッグのデザインを手がけるグラフィックデザイナーは，そう多くはない。コーポレート・アイデンティティのヴィジュアル展開のひとつとしてデザインされたり，デパートなど大手流通業のためのショッピングバッグ・デザインが行なわれてはいるが，大多数を占めるいろいろな店舗のためのショッピングバッグ・デザインは，グラフィックデザイナーの主要な関心事とはなっていないといっていいだろう。それは，多かれ少なかれ，欧米でもあてはまることのようである。ひとつには，ショッピングバッグは，しょせん“使い棄てのチープなバッグ”という現実的な受けとめ方がその背景にあるのかも知れない。
ところが，ショッピングバッグの面白さは，その本質的なマイナス要素をこそ武器とする逆説的な側面にあるといえるのではないだろうか。この本では，東京の原宿・青山・六本木などを重点的に取材して収集した作品を多数掲載しているが，現代日本の最先端の流行と多様な情況をパックしているこの界隈は，世界的にみてもかなりユニークな場所である。ここには，無数のアパレル・ブティックやバラエティ・ショップが集中し，それぞれの独自性をアピールするために，さまざまなショッピングバッグを提供する。バラエティ・ショップはカリカチュア的に，自由奔放に，アパレル・ブティックは目いっぱいのアパレル感覚でショッピングバッグをデザインする。店のオーナーや従業員自らがデザインすることもめずらしくない。
その光景は，使い棄てのショッピングバッグだからこそ，精一杯のメッセージを込めて社会にアピールし，風俗やファッションのなかの一瞬のインパクトになろうと決意しているかのように見える。この独特の立体デザインの一ジャンルにおいて，こうしたグラフィカルでホットな情況が始まったのは，日本ではほんの５，６年前からのこととされているが，実は，そのかなり以前から，こうした，主にアパレル業界の人たちの手によって，さまざまな挑戦が繰り返されてきた実情がある。ファッション界特有のテンポの速さと情況を読みとるすばやさによって，たとえば，あるブティックではシーズンごとにそのデザインを変えるなど，ショッピングバッグの可能性がためされてきたのである。そして，その一方に，そうした多様でめまぐるしい要望に対する，パッケージやショッピングバッグのメーカーの，素材面や機能面，また製造工程での技術的対応があることも忘れてはならないだろう。時代に合ったバッグの姿が追求されるなかで，素材においては紙からビニール，そして安価なポリエチレンやポリプロピレンの比重が増した。機能面でも，把手(とって)や紐に工夫が加えられ，腕に通したり，背負ったりできるさまざまな使い方が生まれてきた。こうして，今，ショッピングやサービスにかかわる重要なツールとして，ショッピングバッグは実に多様な姿を見せているが，そのデザインの自由闊達さは，グラフィックデザイナーを驚かせるに足るものがあるはずである。

この本には，日本全国から募集したものに，東京や関西・中部などの特別取材による作品と世界各国のデザイナーに呼びかけて収集したおよそ3,000点の作品のうち，450点ほどを収録してある。包装紙のあるものは，可能な限りショッピングバッグと一緒に紹介するようにした。

最後に，作品募集に応じていただいた方々，取材に御協力いただいた方々など，たくさんの方々の御支援に対し心から感謝申し上げるとともに，この本が，できるだけ多くの人々に役立ちうるよう，祈ってやまない。

グラフィック社編集部　S.O.

Introduction

The first person to bring a shopping bag to my country high school was a handsome fellow, one whose uniform slacks were tailored to fit nicely. Along with his mandatory black school bag he breezed in that day carrying a white shopping bag. It had apparently come from an apparel store ; what it contained was his gym suit.
This white shopping bag made quite an impression on my classmates ; in fact word of the bag quickly spread throughout the entire school. Undoubtedly someone from his family had carried it back from one of the big cities, and it did have a classy air about it. But more than that it was something that the school had never seen before—it was surprising because it was so totally new.
Few decades ago Japan was just beginning to experience high economic prosperity, and this same phenomenon—a kid stylishly breaking the rules—was undoubtedly occuring in schools all over the country.
The revolution in merchandising has now put major shopping centers in every area of the country. At the same time the transformation in retalling has led to a casting off of the previous concept of what packaging, which is closely related to retailing in the first place, means. The student who carried a shopping bag to school for the first time could not have conceived of the innovation he was playing a part in.

The shopping bag is first of all functional, a tool of modern consumer commerce. It is also a guarantee of service from the enterprise of business to the customer, and an item that must be beautiful and fun to carry home. Its design is vital because its impact on the public is of such importance.
Visual effectiveness is everything. The shopping bag has been called 'walking advertising', a medium that carries the businesses' logo and mark, and thus communicates to the public. For this reason alone great importance must be placed on the design.
Visually—oriented shopping bags began appearing in America and Europe not more than fifteen or twenty years ago. Soon, in the world surrounding Tokyo's Roppongi or Harajuku, fashionable people—models and designers—were to be seen carrying brilliant and psychedelic shopping bags from New York boutiques or London department stores. These flashy, fresh symbols of foreign culture caught the eyes of Tokyo fashion and graphic designers, and left a lasting impression on their artistic sensibilities.
Actually it took a little time for the new concept of shopping bags to take hold in this country, and thoughtful shopping bag design is still not common a thing. As a corporate identity application, and in areas of the major retailing industry, the shopping bag is being approached creatively, but for the most part most graphic designers working for stores do not pay great care to shopping bag design.

This may be true in America and Europe as well. There are still many individuals in business who think of the bag as something cheap, to be thrown away.
In fact this notion—that shopping bags are to be thrown away—is partly true, and can even be used to explain some of the strengths of the shopping bag. Many of the examples in the book were collected from apparel shops and stores in Harajuku, Roppongi and Aoyama, locations on the leading edge of Japan's fashion retailing world. With countless sellers to choose from it is vital that each store projects a vivid image, an idea that sets it apart from the competition. The shopping bag provides just this opportunity. Often in the rush to get new messages or graphic ideas across to patrons, bags for fashion boutiques are being designed with the same speed that new fashions are emerging. And it isn't rare to see these bags designed by store employees, rather than true designers. These creations seemed to be designed with a single thought, to make one brief but highly effective pitch to the public. After achieving that, they may as well be thrown away.
In Japan the excitement that gave rise to this genre only took hold some five or six years ago, but in fashion retailing attempts to use bags creatively probably began much earlier. At some point boutiques began to explore the possibilities of redesigning bags to match the changing of seasonal apparel. In response to this bag and wrapping paper manufacturers began to expand their range of materials and production capabilities. Now makers are moving from paper to the cheaper materials of polyethylene and polypropylene to meet demands for less expensive products. Handles are being improved and bags that can be carried under the arm or over the shoulder are now being designed.
From the standpoint of service and shopping the uses of the shopping bag are obviously great, but the new diversity of designs has surely been a suprise to graphic designers.

For this book contributions were accepted from all over Japan; requests for entries were submitted to designers from around the world as well. Of a total of 3,000 submitted works, some 450 examples were selected. As far as possible wrapping paper was introduced along with shopping bags.

This collection was completed with the kind assistance of contributors, and the support of other individuals too numerous to mention. Along with the greatest thanks to all these people, here's hoping that book provides an inspiration to all the people who open it.

S.O. Editor

凡例

本書には，日本全国から募集した作品に，東京，関西，中部などの特別取材による作品と，世界各国のデザイナーに呼びかけて収集したおよそ3,000点の作品のうち，ヴィジュアル面を重視する編集方針に従って選出したおよそ450点のショッピングバッグ作品を収録してある。包装紙のあるものは，可能な限りショッピングバッグと一緒に紹介した。分類およびレイアウトは，ほぼ，ヴィジュアルのモチーフまた傾向ごとに並べるようにした。そのため，同じ会社や店のショッピングバッグであっても，同一ページに掲載していない場合もある。
作品データは，以下のような順序で示してある。このうち，材料に関しては，紙である場合のみ省略した。

23 ………… 作品・図版番号
ファッション・ブティック ………… 作品タイトル
ポリエチレン，ビニール ………… 材料
ハニー・ブティック ………… 社名・店名
アメリカ，ニューヨーク ………… 所在地
1985 ………… 制作年

AD／アンディ・クレイン ………… アート・ディレクター
D／スコット・ミラー ………… デザイナー
I／ケイト・ミラー ………… イラストレーター
P／ポール・モリシタ ………… フォトグラファー
DF／クレイン・スタジオ ………… デザイン・制作会社
A／US・エージェンシー ………… 代理店

Editorial Notes

This book is a collection of approximately 450 works for shopping bags selected out of some 3,000 works including those sent in response to applications, those resulting from special bag-finding tours in Tokyo, Osaka, Kyoto and Nagoya, and those designed by world-famous designers upon request. The 450 shopping bags shown here were selected in accordance with the editorial policy emphasizing visual aspects of the design, and those with wrapping paper have been shown along with the shopping bag wherever this has been possible. The classifications and the layout have been largely organized according to the visual motif exhibited and so bags of the same company or shop are not necessarily shown on the same page.
The data for the works is indicated in order, as follows. The material is only indicated if it is not paper.

23 ………… Plate Number
Fashion Boutique ………… Title of Work
Plastic ………… Material
Honey Boutique Inc. ………… Client
New York, USA ………… Location of Client
1985 ………… Production Date

AD／Andy Crane ………… Art Director
D／Scott Mirror ………… Designer
I／Kate Mirror ………… Illustrator
P／Pole Morishita ………… Photographer
DF／Crane Studio ………… Design Firm
A／US Agency ………… Agency

SHOPPING BAGS & WRAPPING PAPER

1

1
洋酒会社のギフト用バッグ
サントリー
大阪
1987

D／藤田 隆
DF／サントリー・デザイン部

1
Liquor Company Gift Bag
Suntory Limited
Osaka, Japan
1987

D／Takashi Fujita
DF／Suntory Limited

2
ファッション・ブティック
イッセイミヤケインターナショナル
東京
1984

D／葛西 薫

3
銀行
チェース・マンハッタン
アメリカ, ニューヨーク
1985

AD／ピーター・ハリソン
D／スーザン・ホックバウム
DF／ペンタグラム・デザイン
チェース・マンハッタン銀行のマーケティング・アイデンティティ・プログラムは，国際性のアピールが主眼。これはそのプログラムに従って制作されたショッピングバッグ。強烈なカラーリングとユニークなタイポグラフィーによって，国際性を表現している。

2
Fashion Boutique
Issey Miyake International Inc.
Tokyo, Japan
1984

D／Kaoru Kasai

3
Bank
Chase Manhattan Bank
New York, USA
1985

AD／Peter Harrison
D／Susan Hochbaum
DF／Pentagram Design
Part of a marketing identity program for Chase Manhattan Bank, the brief was to give the design an "international look". Strong primary colors and an unusual typographic approach achieve this goal.

2

85
CHASE
S.I.B.O.S.
PARTNERSHIP IN ACTION

4

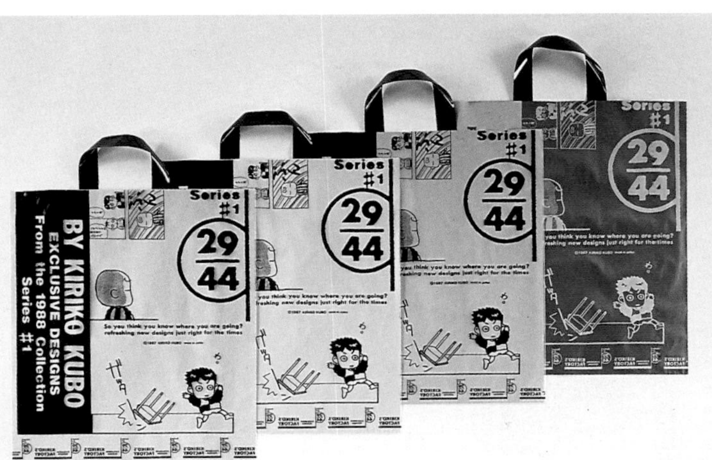

5

6

4・5・6
市販用"キリコズ・ファクトリー"シリーズ
ソニー・クリエイティブ・プロダクツ
東京
1987

AD／信木庸子
D／二瓶光代
I／玖保キリコ
DF／ソニー・クリエイティブ・プロダクツ
玖保キリコのコミックの世界をファッション商品化したもの。未ざらしクラフトのナチュラルな素材感で，キャラクターの個性をストレートに生かそうとしている。

4・5・6
Shopping Bag Series "Kiriko's Factory"
Sony Creative Products Inc.
Tokyo, Japan
1987

AD／Yoko Nobuki
D／Mitsuyo Nihei
I／Kiriko Kubo
DF／Sony Creative Products Inc.
The series of "Kiriko Kubo's Comic World" is regarded as a fashion. This product directly emphasizes the characteristics of the character by taking advantage of the feeling of a material natural to the touch and unbleached kraft pulp.

7

7
皮革品専門店
レザーハウスFree Hand
東京
1977

D／藤田和久
I／藤田千代子
DF／レザーハウスFree Hand
バッグやシューズなど皮革製品のメーカー直営店として，情況に左右されず高品質商品と提供し続ける姿勢を表現するものとして，孤高な“狼”の姿をシンボルに使っている。このシンボルは，ショッピングバッグのほか，パッケージ類にも使用されている。

7
Leather Goods Boutique
Free Hand
Tokyo, Japan
1977

D／Kazuhisa Fujita
I／Chiyoko Fujita
DF／Free Hand
This leather goods maker sells high quality leather shoes and bags through its own store. The firm's symbol is the lone wolf, which represents unswerving commitment to quality and longstanding business. It appears on shopping bags and various packages.

8

8
ファッションビル
原宿ル・ポンテ
東京
1983

AD／湯村輝彦
D／藤原哲男
I／湯村輝彦, 湯村タラ
DF／フラミンゴ・スタジオ, ステレオスタジオ
A／サザンプトン

9
デパートの父の日キャンペーン
三越
東京
1987

AD／安藤 亨
D／木本映右
I／田中紀之
DF／三越デザイン開発室
A／博報堂

8
Boutique Shopping House
Harajuku Le Ponte
Tokyo, Japan
1983

AD／Teruhiko Yumura
D／Tetsuo Fujiwara
I／Teruhiko Yumura, Tara Yumura
DF／Flamingo Studio Inc, Stereo Studio Inc.
A／Southampton

9
Department Store Father's Day Campaign
Mitsukoshi Ltd.
Tokyo, Japan
1987

AD／Tohru Ando
D／Eisuke Kimoto
I／Noriyuki Tanaka
DF／Design Room, Mitsukoshi Ltd.
A／Hakuhodo, Inc.

9

10

11

12

10
テナントビルの夏のキャンペーン
岡田屋モアーズ
横浜・川崎
1985

AD／杉本英介
D／永井裕明
P／若月 勤
DF／ブレックファースト

11
テナントビルの冬のキャンペーン
岡田屋モアーズ
横浜・川崎
1984

AD／杉本英介
D／永井裕明
I／飯田三代
DF／ブレックファースト

12
テナントビル冬のキャンペーン
岡田屋モアーズ
横浜・川崎
1987

AD／杉本英介
D／永井裕明
I／大西重成
DF／ブレックファースト

13
ファッションビル
ラフォーレ原宿
東京

AD／ラフォーレ原宿

10
Boutique Shopping House
Summer Campaign
Okadaya-More's
Yokohama, Japan
1985

AD／Eisuke Sugimoto
D／Hiroaki Nagai
P／Tsutomu Wakatsuki
DF／Breakfast Co., Ltd.

11
Boutique Shopping House
Winter Campaign
Okadaya-More's
Yokohama, Japan
1984

AD／Eisuke Sugimoto
D／Hiroaki Nagai
I／Miyo Iida
DF／Breakfast Co., Ltd.

12
Boutique Shopping House
Winter Campaign
Okadaya-More's
Yokohama, Japan
1987

AD／Eisuke Sugimoto
D／Hiroaki Nagai
I／Shigenari Onishi
DF／Breakfast Co., Ltd.

13
Boutique Shopping House
Laforet Harajuku
Tokyo, Japan

AD／Laforet Harajuku

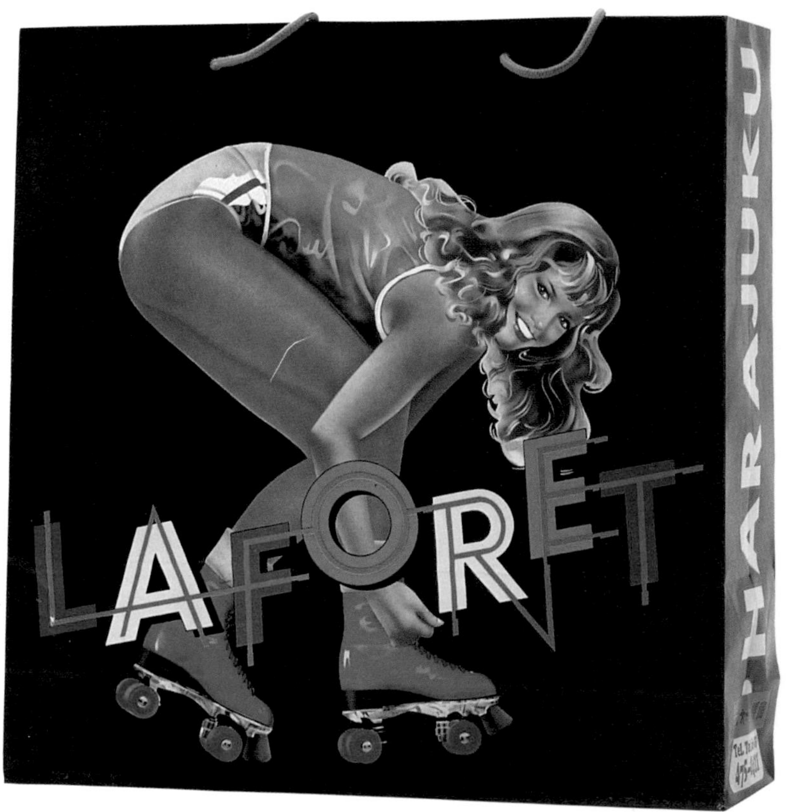

13

14
ファッションビル
ラフォーレ原宿
東京
1986

D／永島 圭
I／デービッド・ピータース
DF／USSO(ウーゾ)

14
Boutique Shopping House
Laforet Harajuku
Tokyo, Japan
1986

D／Kei Nagashima
I／David Peters
DF／USSO

14 a

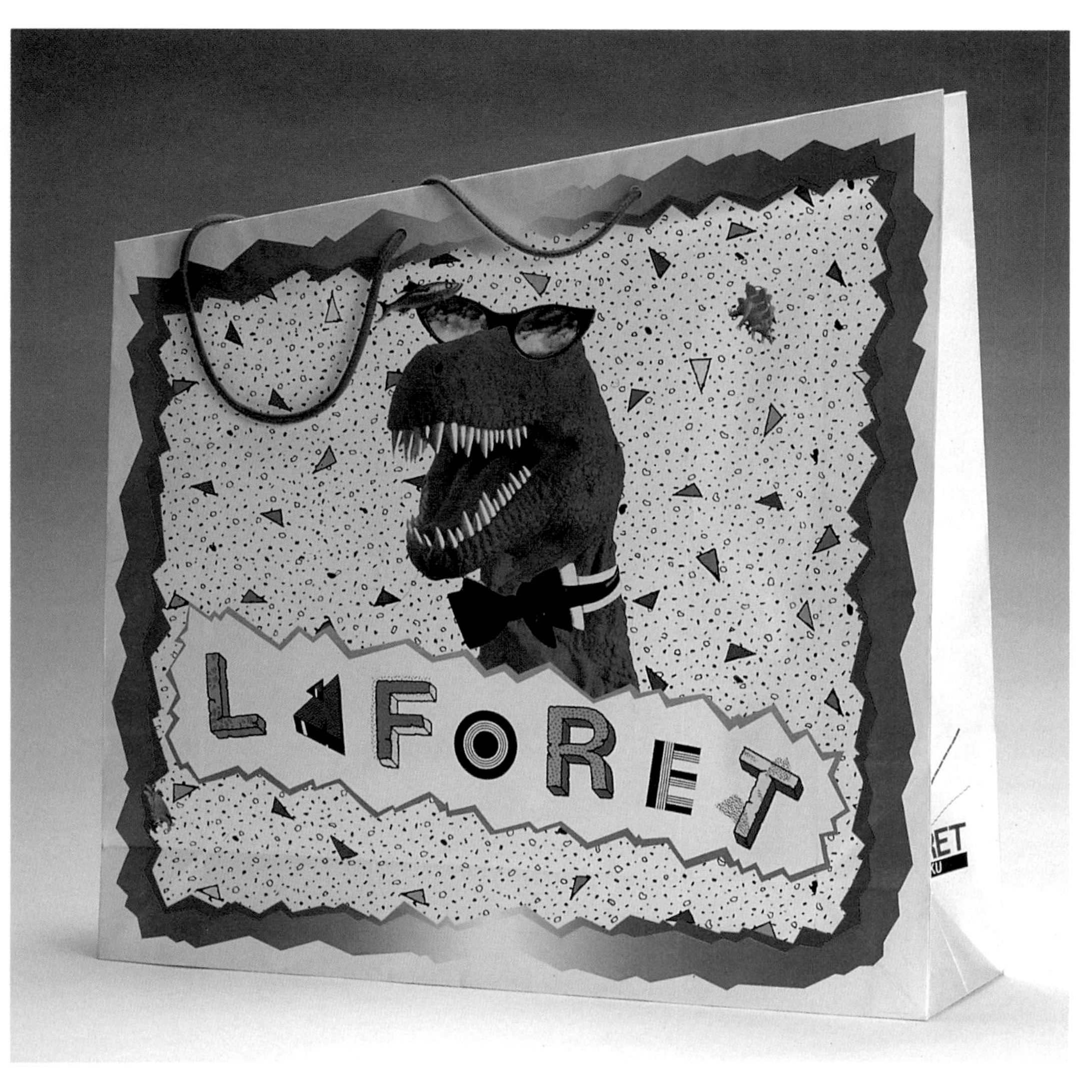

14 b

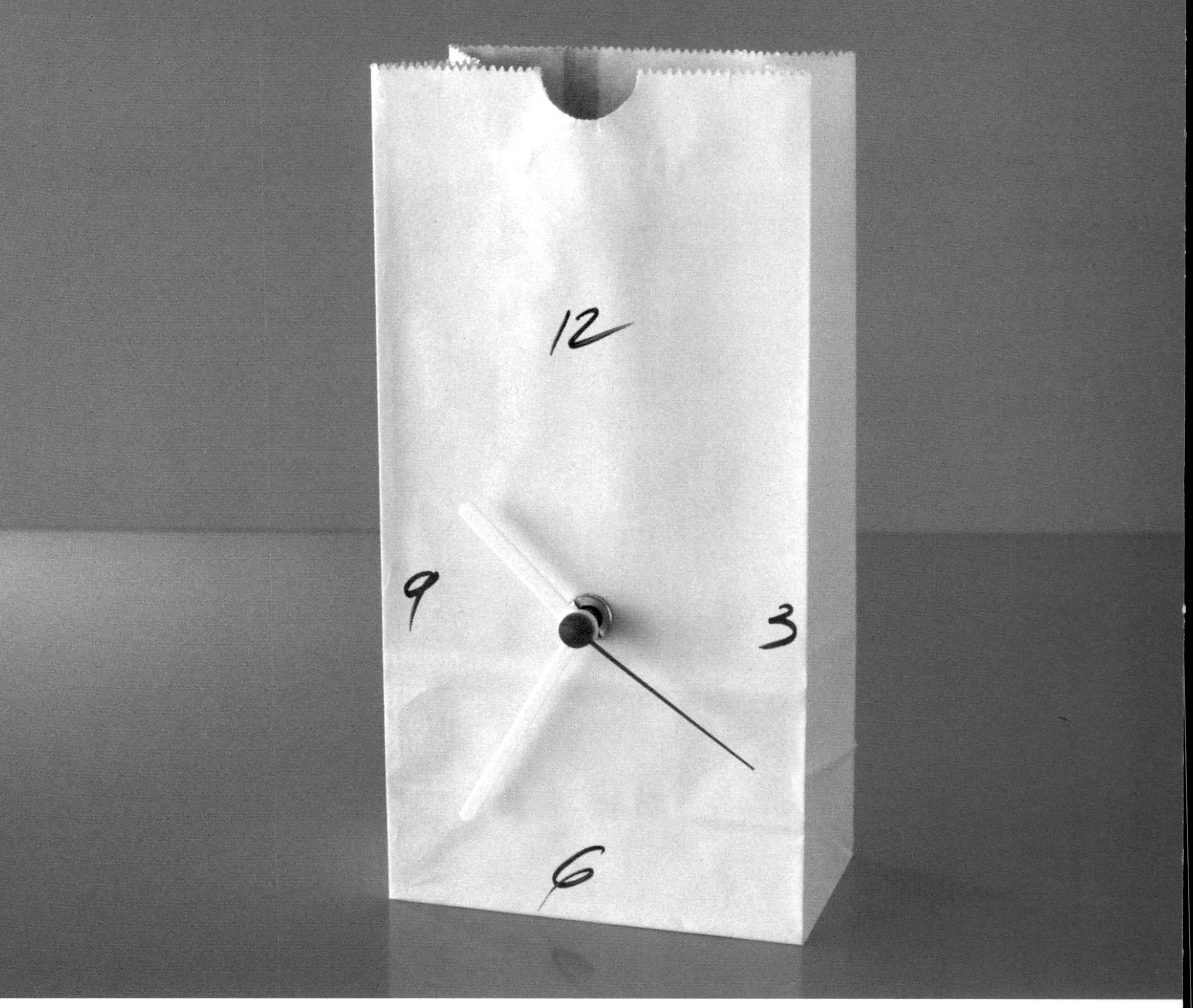

15

15
"ペーパーバッグ・クロック"
パワー・ハウス・ミュージアム
オーストラリア, シドニー
1988

D/フェルナンド・メディーナ
DF/メディーナ・デザイン
オーストラリア・シドニーの"パワー・ハウス・ミュージアム"に永久収蔵されている"ペーパーバッグ・クロック"(紙袋の時計)。

15
"Paper Bag Clock"
Power House Museum
Sydney, Australia
1988

D/Fernando Medina
DF/Medina Design
This "Paper Bag Clock" is included in the permanent collection of the Power House Museum of Sydney, Australia.

16
"ペパーミント・ジェリー"
ペパーミント・パラレル
オランダ, アムステルダム
1977

D／アントン・ビーケ
P／チールド・フレデリックス
DF／アントン・ビーケ・アンド・アソシエイツ

17
ブティック
Dourakuya Grp.（ディ・ショップ）
東京
1986

D／ダンハイツ企画部（メンズ）, D_2 コレクション企画部（レディース）

16
"Peppermint Jelly"
Peppermint Parallel
Amsterdam, The Netherlands
1977

D／Anthon Beeke
P／Tjeerd Frederiks
DF／Anthon Beeke & Assoc.

17
Boutique
Dourakuya Grp. (D.Shop Co., Ltd.)
Tokyo, Japan
1986

D／Dan-Haitu Planning Dept. (for men), D_2 Collection Planning Dept. (for ladies)

16

17

18
民芸品店
ブルー＆ホワイト
東京
1983

D／加藤エイミー，高橋由紀子

19
個展のためのショッピングバッグ
東京デザイナーズスペース
東京
1987

D／川崎修司，高橋新三
P／能津喜代房
DF／ヘッズ

18

19

18
Souvenir Shop
Blue & White
Tokyo, Japan
1983

D／Eimy Kato, Yukiko Takahashi

19
Shopping Bag Works for the One-man Exhibit
Tokyo Designer's Space
Tokyo, Japan
1987

D／Shuji Kawasaki, Shinzo Takahashi
P／Kiyofusa Nozu
DF／Heads Inc.

20
デパートのオリジナル風呂敷
布
松屋
東京
1986

AD／水野皓司
D／舟橋全二
DF／松屋
デパートのオリジナル風呂敷で，春夏秋冬それぞれのモチーフで４種類つくられたもの。椿の花のデザインは春。柿の実のモチーフは秋。

21
ファッションビル
Privy
札幌
1985

AD／斉正隆男
I／舟橋全二
DF／ノルド

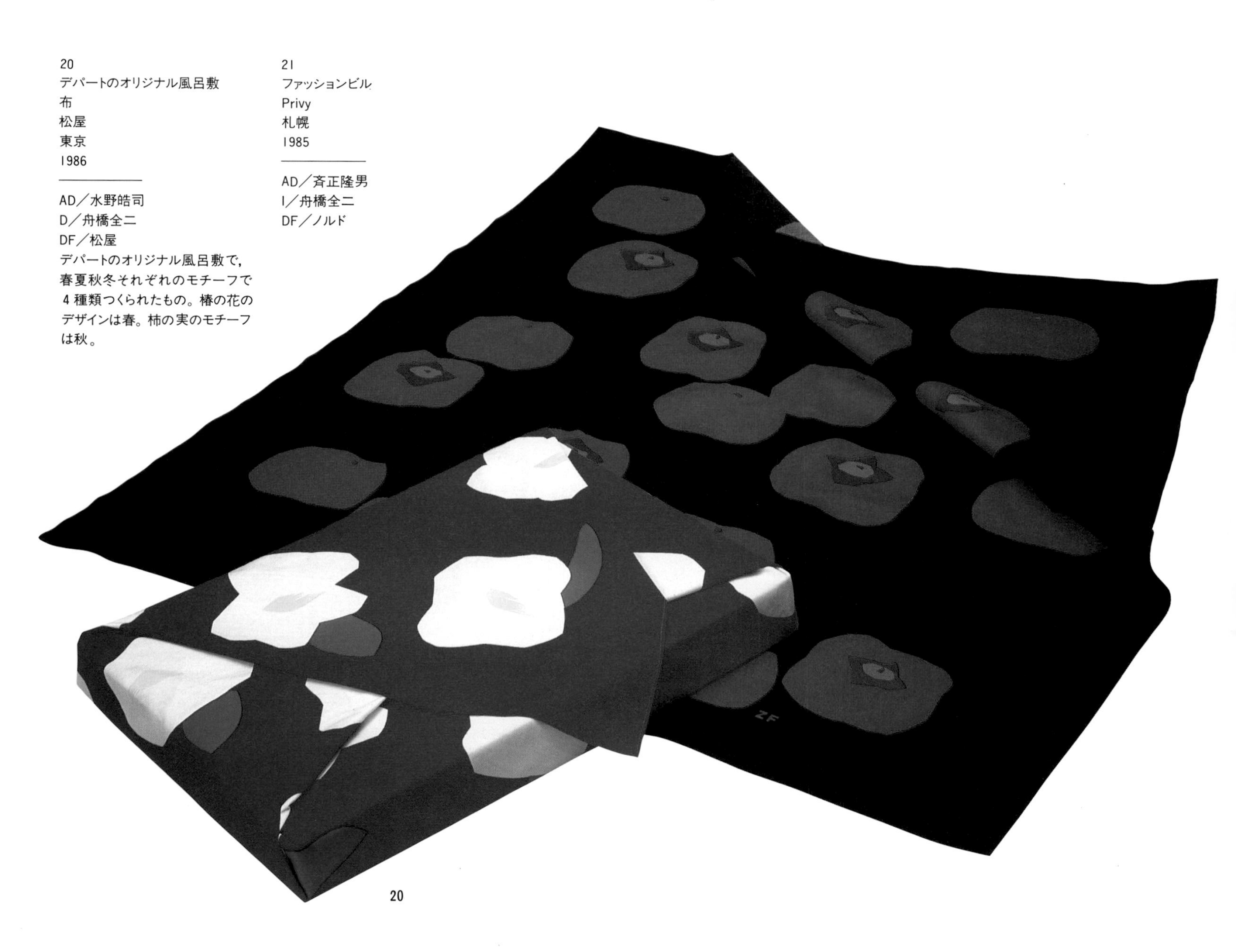

20

21

20
Department Store Original Wrapping Cloth
Matsuya Co., Ltd.
Tokyo, Japan
1986

AD／Koji Mizuno
D／Zenji Funabashi
DF／Matsuya Co., Ltd.
Furoshiki (wrapping cloth) designed for each of the seasons of spring, summer, autumn and winter in the original brand by a department store. The one for spring demonstrates camellias. The one with the persimmon motif is for autumn.

21
Shopping Center
Privy
Sapporo, Japan
1985

AD／Takao Saijo
D／Zenji Funabashi
DF／Nord Co., Ltd.

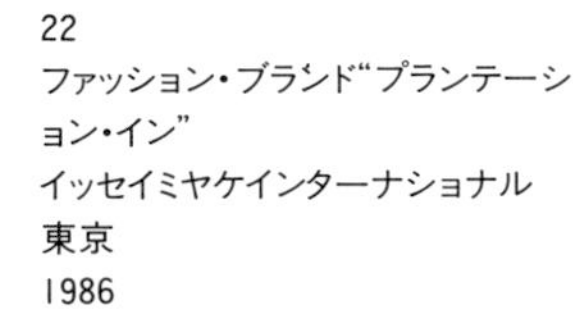

22
ファッション・ブランド"プランテーション・イン"
イッセイミヤケインターナショナル
東京
1986

AD／成瀬始子
D／成瀬始子, 河上妙子
原型は, 低ロット・低コストを考慮した紙製の平袋。洋服を入れると自然にふくらみができるが, そのふくらみを利用して折るとテトラ型のリュックサックが出来上がる。背負うことができる。

24
ファッション・ブランド"プランテーション・イン"
ビニール
イッセイミヤケインターナショナル
東京
1986

AD／成瀬始子
D／成瀬始子
河上妙子

22
Fashion Brand "Plantation In"
Issey Miyake International Inc.
Tokyo, Japan
1986

AD／Motoko Naruse
D／Motoko Naruse, Taeko Kawakami
A long flat paper bag, designed taking into consideration low lot and low cost. When clothes are placed inside, it naturally fills out and one can make a tetrad shape ruck sack if one then folds the bag. This bag can be carried on one's back.

24
Fashion Brand "Plantation In"
Plastic
Issey Miyake International Inc.
Tokyo, Japan
1986

AD／Motoko Naruse
D／Moroko Naruse, Taeko Kawakami

22

24

23

23
ファッション・ブティック
イッセイミヤケインターナショナル
東京
1987

D／津森千里
DF／イッセイミヤケインターナショナル

25
ブティック
ジョイス・ブティック
ホンコン
1981

D／ヘンリー・スタイナー
DF／グラフィック・コミュニケーション

26
ブティック
ベーリー・ストックマン
東京
1986

AD／井上義朗
D／草野和芳
DF／ル・ギャマン

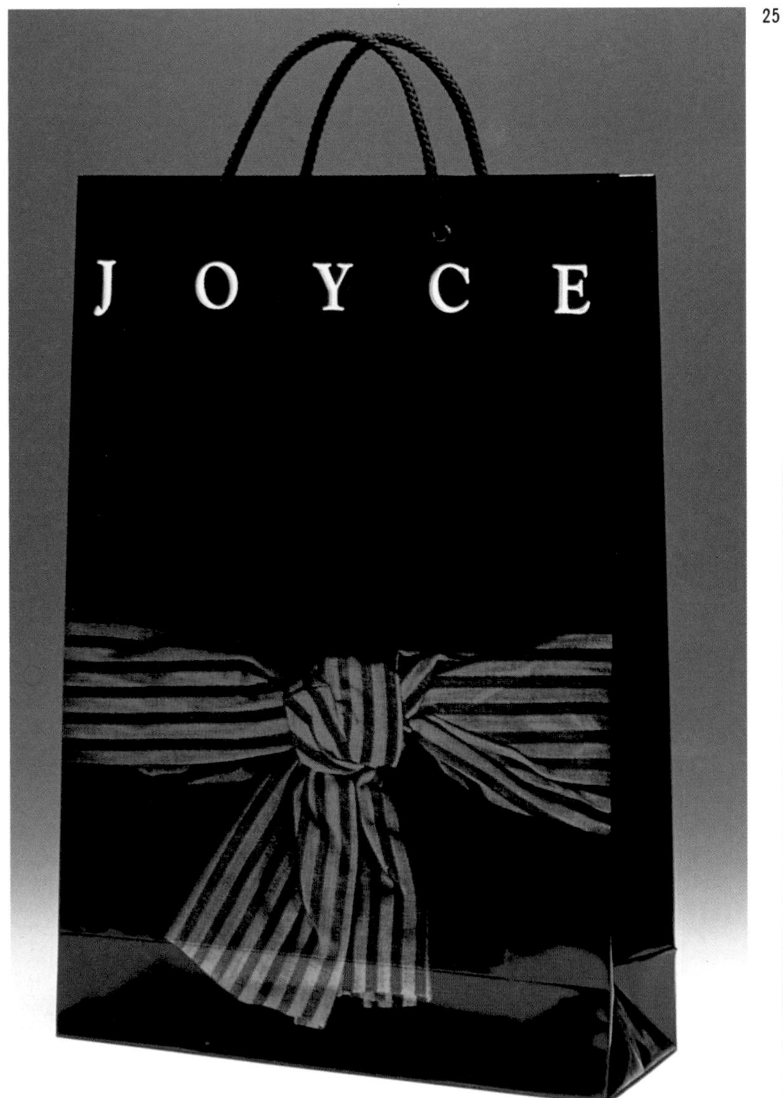

25

23
Fashion Boutique
Issey Miyake International Inc.
Tokyo, Japan
1987

D／Chisato Tsumori
DF／Issey Miyake International Inc.

25
Boutique
Joyce Boutique Ltd.
Hong Kong
1981

D／Henry Steiner
DF／Graphic Communication Ltd.

26
Boutique
Bailey Stockman Co., Ltd.
Tokyo, Japan
1986

AD／Yoshiro Inoue
D／Kazuyoshi Kusano
DF／Le Gamin

26

27

27
ミュージアム・ショップ
ビニール
ニューヨーク近代美術館
アメリカ, ニューヨーク
1988

D／五十嵐威暢
DF／イガラシステュディオ

27
Museum Shop
Plastic
The Museum of Modern Art, New York
New York, USA
1988

D／Takenobu Igarashi
DF／Igarashi Studio

28

28
ショッピング・モール"ガレリア"
ジェラルド・D・ハイネス・インタレスツ
アメリカ, テキサス州ダラス
1982

D/ウッディ・パートル
DF/パートル・デザイン

28
Shopping Mall "Galleria"
Gerald D. Hines Interests
Dallas, Tex., USA
1982

D/Woody Pirtle
DF/Pirtle Design

29

30

31

29
テナントビルの夏のキャンペーン
岡田屋モアーズ
横浜・川崎
1986

AD／杉本英介
D／永井裕明
DF／ブレックファースト

30
アパレル・ブティック
ライカ（ピアージュ）
大阪
1985

D／佐々木 操

31
美術館
ソロモン・R・グッゲンハイム美術館
アメリカ，ニューヨーク
1984

D／マルコム・グレア・デザイナーズ・インコーポレイテッド
このパターンを使ったイメージは，有名なフランク・ロイド・ライトの建築にもとづいてデザインされた。

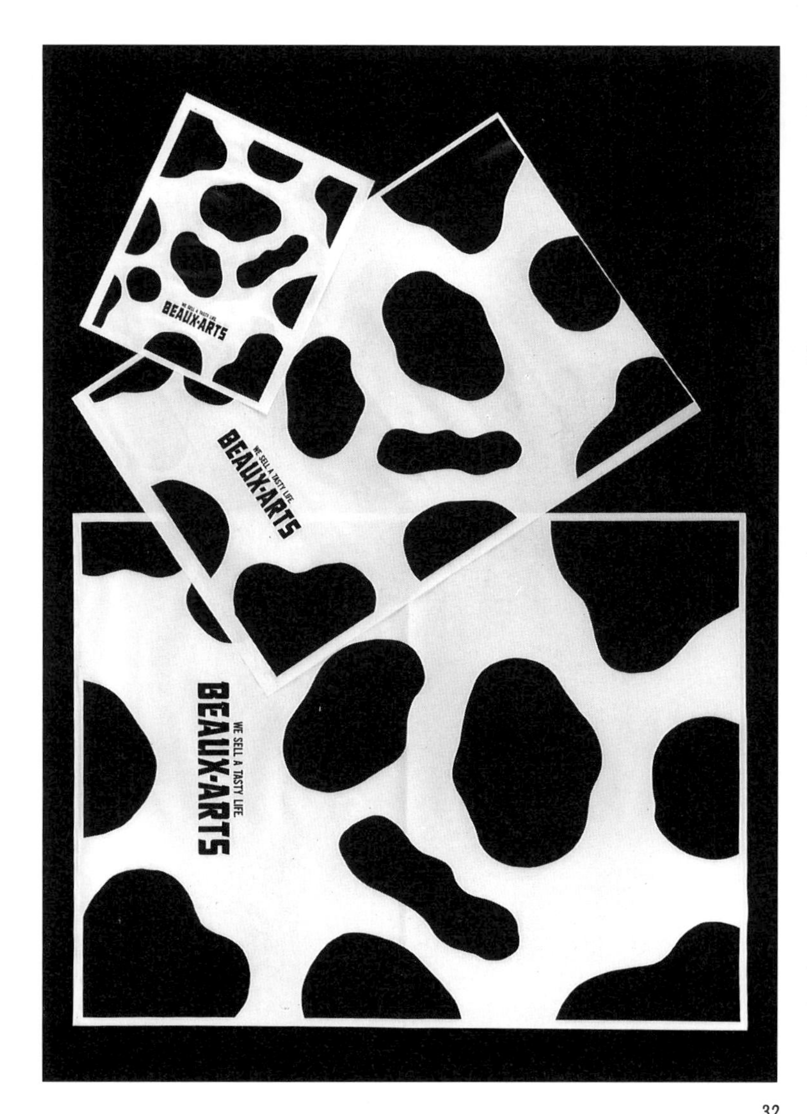

32

33

32
雑貨店
ビニール
ブルーグラス(ボザール)
東京
1985

D/遠藤陽子

33
美術館
ソロモン・R・グッゲンハイム美術館
アメリカ, ニューヨーク
1975

D/マルコム・グレア・デザイナーズ・インコーポレイテッド
フランク・ロイド・ライトの円形ランプの形をもとにデザインされたショッピングバッグ。

29
Boutique Shopping House
Summer Campaign
Okadaya-More's
Yokohama, Japan
1986

AD/Eisuke Sugimoto
D/Hiroaki Nagai
DF/Breakfast Co., Ltd.

30
Apparel Boutique
Raika Co., Ltd. (Piage)
Osaka, Japan
1985

D/Misao Sasaki

31
Museum
Solomon R. Guggenheim Museum

New York, USA
1984

D/Malcolm Grear Designers, Inc.
The image used for the pattern is based on the well-known Frank Lloyd Wright building.

32
Variety Shop
Plastic
Blue Glass (Beaux-arts)
Tokyo, Japan
1985

D/Yoko Endo

33
Museum
Solomon R. Guggenheim Museum
New York, USA
1975

D/Malcolm Grear Designers, Inc.
The design is based on the shapes of the circular ramps within the Frank Lloyd Wright building.

34

34
店舗総合見本市"ジャパン・ショップ"
日本経済新聞社
東京
1987

D／松永 真
日本経済新聞の主催で開かれる国際的な店舗総合見本市のためのショッピングバッグ。1987年に「店と街，いま，いきいき宣言」というスローガンで開催されたときのメーンパターンで，人間をかたちどったもの。

35
複合ビル"WALTZ"
所沢市
埼玉県所沢市
1986

D／松永 真
"WALTZ"(ワルツ)は，所沢市の商業開発ビルの名称。デパートの"所沢西武"がそのなかに開設された。デパートのオープニング・キャンペーン用ショッピングバッグとともに，WALTZ 自体のショッピングバッグも制作されたもので，ロゴそのものが全体的なモチーフになって大懸垂幕などにも展開された。

34
Comprehensive Store Exhibition "Japan Shop"
Nihon Keizai Shimbun, Inc.
Tokyo, Japan
1987

D／Shin Matsunaga
The comprehensive store exhibition "Japan Shop", is an international exhibition held at Harumi every year under the auspices of the Nihon Keizai Shimbun Newspaper. This pattern, which stands for the human body, was the main pattern of the 1987 exhibition, with the slogan "Declaration for the healthy and lively growth of the store and the town".

35
Building Complex "WALTZ"
Tokorozawa City
Tokorozawa City, Japan
1986

D／Shin Matsunaga
"Waltz" is the nickname of the Tokorozawa Municipal Building Complex. The Tokorozawa Seibu Department Store opened in this building as a key tenant. At the opening ceremony, the logo itself was used as an overall motif and was presented on the giant banners and shopping bags.

35

36
果物店
銀座千疋屋
東京
1965

D／橋岡庸晴

37
美術館の50周年記念祭
ニューヨーク近代美術館
アメリカ, ニューヨーク
1979

D／トム・ガイスマー
DF／シャマイエフ・アンド・ガイスマー・アソシエイツ

38
"ジュロン・バード・パーク"
ジュロン・バード・パーク
シンガポール
1987

D／ケン・ケイトー
DF／ケイトー・デザイン・インコーポレイテッド

36
Fruit Shop
Ginza Senbikiya
Tokyo, Japan
1965

D／Tsuneharu Hashioka

37
The 50th Anniversary of the Museum of Modern Art
The Museum of Modern Art, New York
New York, USA
1979

D／Tom Geismar
DF／Chermayeff & Geismar Associates

38
"Jurong Bird Park"
Jurong Bird Park
Singapore
1987

D／Ken Cato
DF／Cato Design Inc.

36

37

38

39
ファッション・ブランド
"Junko Shimada Part 2"
ルシアン・プランニング
東京
1987

CD／加来憲司
AD／矢野邦彦
D／南 百合江
DF／C・カンパニー
ファッション・ブランドのスポーティで遊び心のある雰囲気，機能性を，ロゴマークのバリエーションでショッピングバッグにも展開。

39
Fashion Brand "Junko Shimada Part 2"
Lecien Planning Corporation
Tokyo, Japan
1987

CD／Kenji Kaku
AD／Kunihiko Yano
D／Yurie Minami
DF／C Company Ltd.
Creating an image of practicality and a sporty, leisurely atmosphere, this shopping bag has been designed with a variety of logos.

39

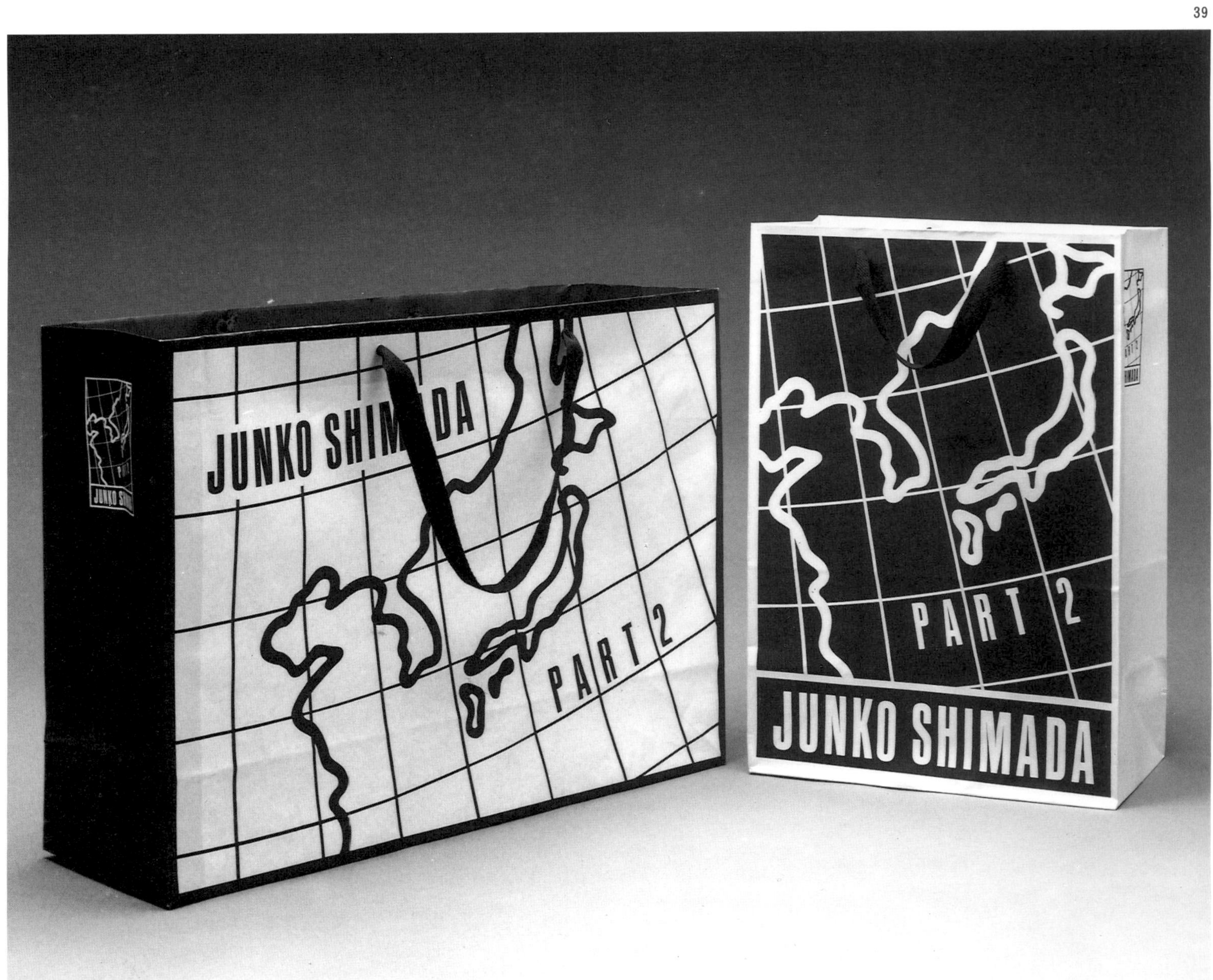

40

41

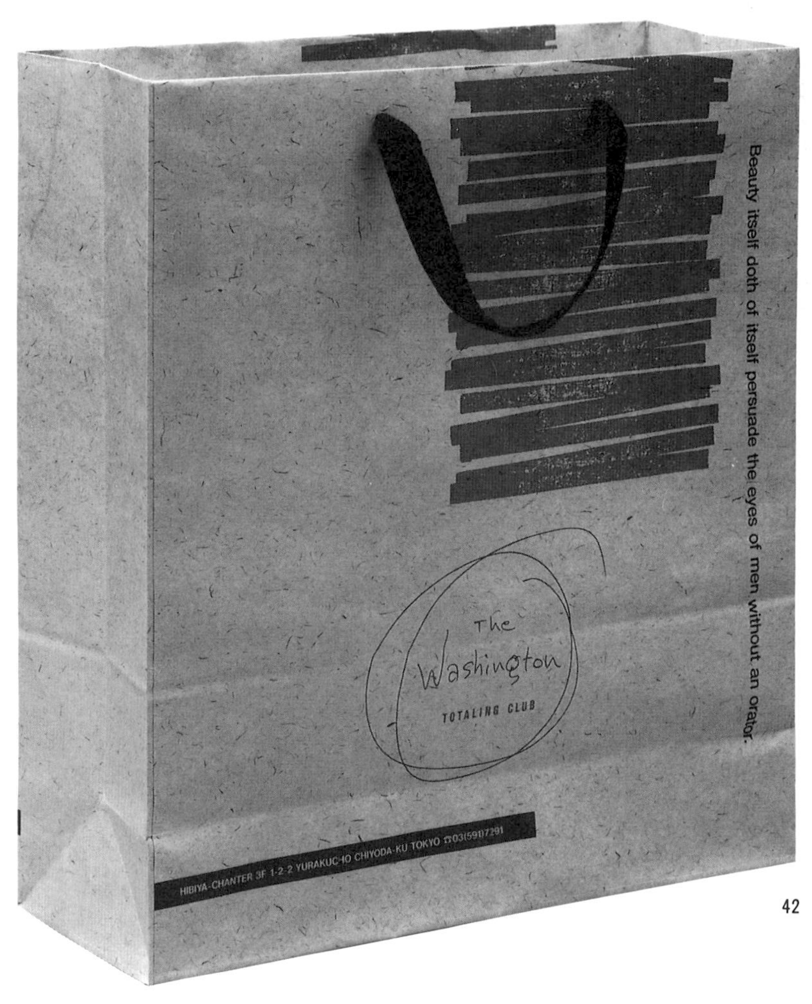

42

40
ファッションビル
大宮ステーションビル
埼玉県大宮市
1987

D／八木健夫
DF／オフィス・ピーアンドシー

41
量販店の新年セール
田原屋
川崎市
1988

D／東京グラフィック

42
靴店
ワシントン靴店
東京
1987

D／角掛晶宜
DF／CA
A／カサッポ＆アソシエイツ
靴を中心としたメンズ・グッズ・ショップのショッピングバッグ。素材はナチュラルなクラフト紙を使用し，グラフィック処理で男の新しいアクティブなイメージを表現している。

40
Shopping Center
Omiya Station Building
Omiya City, Japan
1987

D／Takeo Yagi
DF／Office P & C Inc.

41
Superstore New Year Sale
Tawaraya Co., Ltd.
Kawasaki City, Japan
1988

D／Tokyo Graphic Co., Ltd.

42
Men's Shoes Shop
Washington Shoe Co., Ltd.
Tokyo, Japan
1987

D／Masanobu Tsunokake
DF／CA
A／Casappo & Assoiates
A bag for a shop specializing in men's shoes. The material used is natural kraft paper, illustrating new men's image using a graphic arts process.

43

43
"Sukeru 紙袋"
紙+ワックス
キジュウロウ・ヤハギ
東京
1984

D／矢萩喜従郎
DF／キジュウロウ・ヤハギ

44
ガラスウェア・メーカー
東洋ガラス
東京
1986

AD／片岡　脩
D／金井幸治, 浜 恭子
DF／シウ・グラフィカ

43
"Sukeru (Transparent) Paper Bag"
Paper+Wax
Kijuro Yahagi Co., Ltd.
Tokyo, Japan
1984

D／Kijuro Yahagi

44
Glass Wear Manufacturer
Toyo Glass Co., Ltd.
Tokyo, Japan
1986

AD／Shu Kataoka
D／Koji Kanai, Kyoko Hama
DF／Siu Graphica Inc.

45
化粧品会社の社長就任式用ハンディバッグ
資生堂
東京
1987

AD／中山禮吉, 村井和章
D／村井和章
DF／資生堂宣伝部
資生堂のマークに使われている椿を楕円で表現し, 上部の暗いグラデーションの中に置いたデザイン。昇る太陽がイメージの基調。

46
注文家具フェア
ダラス・マーケット・センター(インフォワークス)
アメリカ, テキサス州ダラス
1984

AD／ウッディ・パートル
D／ウッディ・パートル, ジェフ・ワイスマン
DF／パートル・デザイン
注文に応じて家具のデザインを手掛けているテキサス州のダラス・マーケット・センターのバッグ。このグラフィックスの主眼は"デザイン, テクノロジー, 生産性のためのインフォメーション"というもの。

44

45
Cosmetics Company
Shiseido Co., Ltd.
Tokyo, Japan
1987

AD/Reikichi Nakayama, Kazuaki Murai
D/Kazuaki Murai
DF/Shiseido Co., Ltd.
This design uses the camellia symbol of the Shiseido Cosmetics Company in an elliptical form, with a dark gradation at the upper section. The rising sun forms the central image.

46
Contract Furnishing Exhibition
Dallas Market Center
(Infoworks)
Dallas, Tex., USA
1984

AD/Woody Pirtle
D/Woody Pirtle, Jeff Weithman
DF/Pirtle Design
The bag for a contract design furniture market in Dallas, Texas. The graphics symbolize the theme "Information That Works....Design, Technology, Productivity".

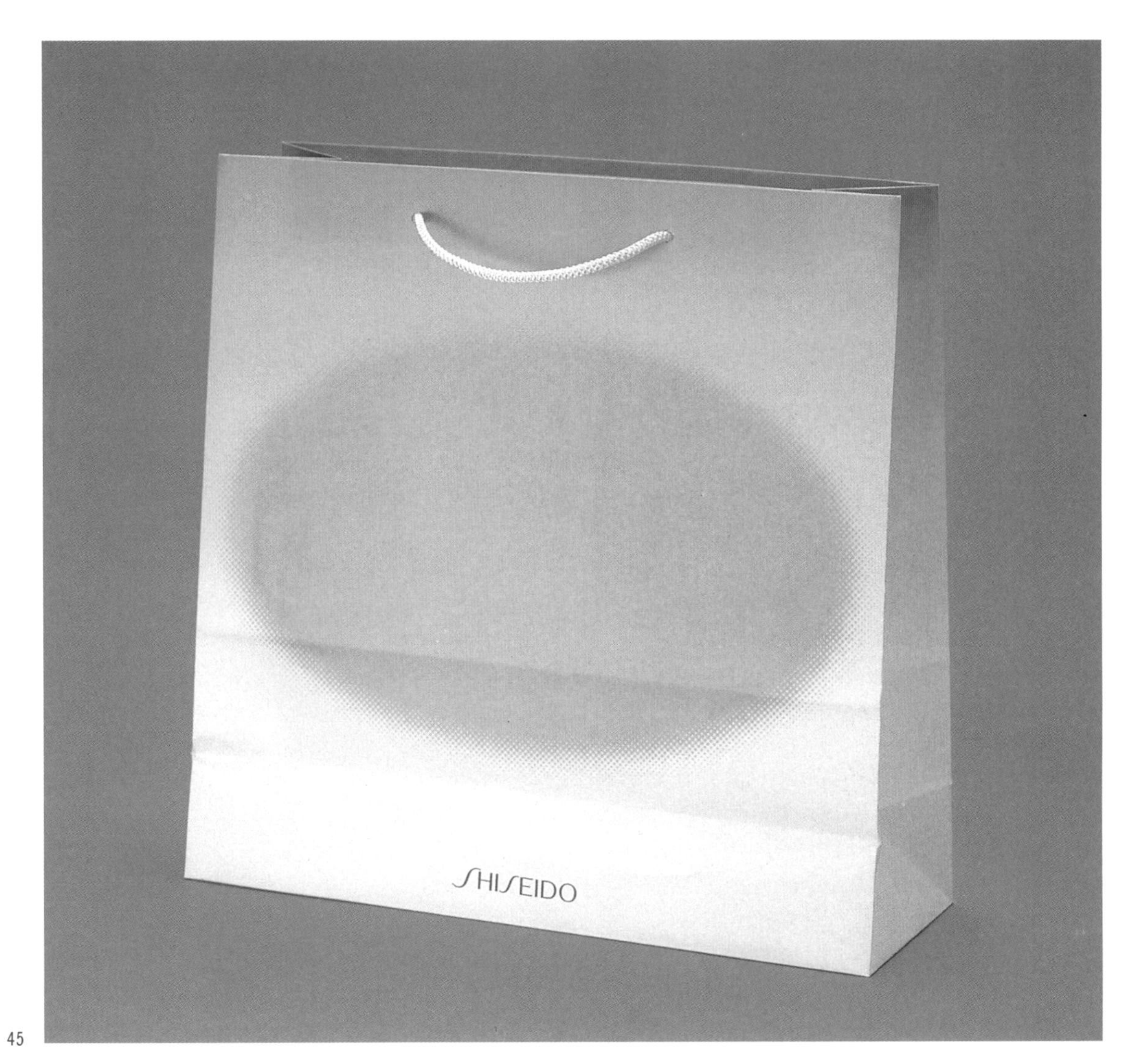

45

46

47

48

47
"センチュリー・シティ・ショッピング・センター"
ザ・リーフ・コーポレーション
アメリカ, ロサンゼルス

D／マイケル・マナリング
DF／オフィス・オブ・マイケル・マナリング

48
スポーツクラブ
ビニール
スポーツパレス・ジスタス
那覇市
1987

AD／久留米 裕
D／本田良治
DF／トライ

49・50
デパート
ブルーミングデイルズ
アメリカ, ニューヨーク
1979

D／ダン・レイジンガー
DF／ストゥディオ・レイジンガー

51
レディース・ブティック
ビニール
イマーゴ
アメリカ, ハワイ

D／エルバート・K・ツチモト
DF／K & E グラフィックス
イマーゴはスポーティなヨーロピアン・カジュアル中心のレディース・ブティック。イマーゴとは毛虫から蝶へ変わることを意味する。

47
"Century City Shopping Center"
The Rreef Corporation
Los Angeles, USA

D／Michael Manwaring
DF／The Office of Michael Manwaring

48
Sports Club
Plastic
Sport Palace Xystus
Naha City, Japan
1987

AD／Yutaka Kurume
D／Yoshiharu Honda
DF／Try Co., Ltd.

49・50
Department Store
Bloomingdale's
New York, USA
1979

D／Dan Reisinger
DF／Studio Reisinger

51
Ladies' Boutique
Plastic
Imago
Hawaii, USA

D／Elbert K. Tsuchimoto
DF／K & E Graphic, Inc.
Imago is a contemporary women's boutique featuring sporty European design clothing. The word "Imago" means the transformation of a caterpillar into a butterfly.

AEL
the dream
ril 30- may 19, 1979.
49

ISRAEL
the dream
ISRAEL
the dream
ISRAEL
the dream
50

IMAGŌ
51

52

53

52
デパート
ブルーミングデイルズ
アメリカ, ニューヨーク
1984

AD／ジョン・ジェイ
D／永井一正

53
デパート
高島屋
東京
1984

D／安部修二
DF／TAD

52
Department Store
Bloomingdale's
New York, USA
1984

AD／John Jay
D／Kazumasa Nagai

53
Department Store
Takashimaya Co., Ltd.
Tokyo, Japan
1984

D／Shuji Abe
DF／TAD Co., Ltd.

54

55

54
ブティック
ビニール
ナイス
東京
1987

D／木村力三
あえて店名を入れず，各国の国旗のもつ面白さをアーティスティックに表現。

54
Boutique
Plastic
Nice & Company
Tokyo, Japan
1987

D／Rikizo Kimura
A shopping bag without the name of the company or shop printed on it, displaying an interesting design of flags from all over the world.

55
キャラクター商品用ショッピングバッグ
サントリー
大阪
1984

AD／山登道雄
D／岡崎美保
DF／サントリー・デザイン部

55
Shopping Bag for Character Products
Suntory Limited
Osaka, Japan
1984

AD／Michio Yamato
D／Miho Okazaki
DF／Suntory Limited

56
"ヘルツ"
ビニール
ヴィクター・メイヤー
スイス，オルテン
1970

D／アーミン・フォークト
DF／アーミン・フォークト・パートナーズ

56
"Herz"
Plastic
Victor Meyer
Olten, Switzerland
1970

D／Armin Vogt
DF／Armin Vogt Partners

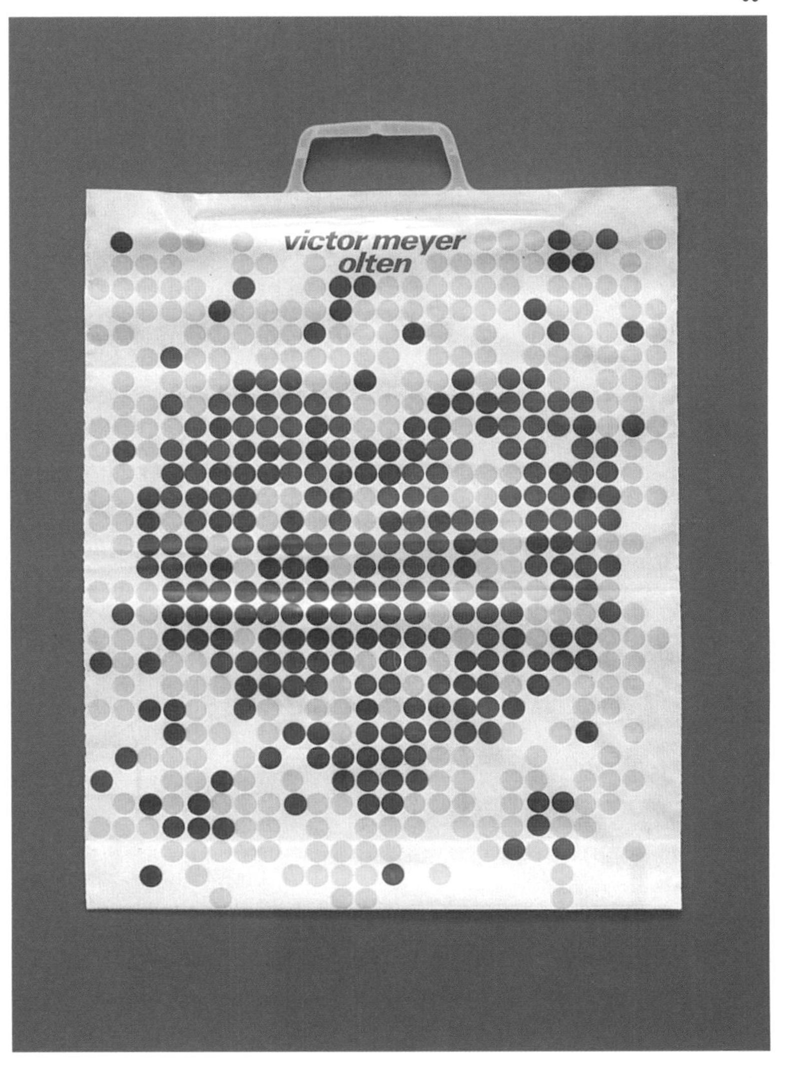

56

57

57
バラエティ・ショップのバレンタイン・デー用とホワイト・デー用ショッピングバッグ
ソニー・プラザ
東京
1987

D／渡辺通正

58
和菓子販売会社
ふるさとや
東京
1986

AD／長友啓典
D／長友啓友, 黒田征太郎, 野村高志
DF／K2

57
Valentine's Day Bag and White Day Bag of Variety Shop
Sony Plaza Co., Ltd.
Tokyo, Japan
1987

D／Michimasa Watanabe

58
Japanese Confectionery
Furusatoya
Tokyo, Japan
1986

AD／Keisuke Nagatomo
D／Keisuke Nagatomo, Seitaro Kuroda, Takashi Nomura
DF／K-Two Co., Ltd.

58

59
キャラクター商品用ショッピングバッグ
サントリー
大阪
1985

AD／山登道雄
D／藤本勝之
DF／サントリー・デザイン部

60
ファッション会社のクリスマス・バザール
オオコシ
東京
1987

AD／神保米雄
D／伊藤周三
DF／マック・ジャパン
A／伊藤紙工
クリスマス・バザール用の販促ツールとしてPOP，シール，ショッピングバッグ，DMなどに展開されたものの一部。

59
Shopping Bag for Character Products
Suntory Limited
Osaka, Japan
1985

AD／Michio Yamato
D／Katsuyuki Fujimoto
DF／Suntory Limited

60
Fashion Company Christmas Bazaar
Ohkoshi Co., Ltd.
Tokyo, Japan
1987

AD／Yoneo Jimbo
D／Shuzu Ito
DF／Mac Japan Co., Ltd.
A／Itoshiko Co., Ltd.
Part of the works developed into the form of POP, seal, shopping bag, and direct mail as sales promotion tools for a Christmas bazaar.

59

60

61
デパート
近鉄百貨店
大阪
1987

AD/田中康夫, 能見統之
D/田中康夫
DF/パッケージ・ランド
A/東洋紙業
近鉄百貨店の"K"をシンボライズしてデザインされた小さめのショッピングバッグ。

62
市販用"ミラノ"
松城紙袋工業
東大阪市
1987

D/田中康夫
DF/パッケージ・ランド

63
"マトリキュラ"
ビニール
マトリキュラ
スペイン, バレンシア
1986

D/ジョルジョ・グレゴリー
DF/アルキミア

64
フランクフルト見本市
メッセ・フランクフルト
西ドイツ, フランクフルト

DF/スタンコウスキー・プラス・デュシェク

65
タイル・住宅設備機器会社
INAX
東京
1985

AD/中西元男
D/パオス

61
Department Store
Kintetsu Department Store Co., Ltd.
Osaka, Japan
1987

AD/Yasuo Tanaka, Toshiyuki Nomi
D/Yasuo Tanaka
DF/Package Land Co., Ltd.
A/Toyoshigyo Co., Ltd.
A design using the "K" symbol of the Kintetsu Department store.

62
Shopping Bag "Milano"
Matsushiro Paper Bag Industry Co., Ltd.
Osaka, Japan
1987

D/Yasuo Tanaka
DF/Package Land Co., Ltd.

63
"Matricula"
Plastic
Matricula
Valencia, Spain
1986

D/Giorgio Gregori
DF/Alchimia

64
"Messe Frankfurt"
Messe Frankfurt GMBH
Frankfurt, West Germany

DF/Stankowski+Dushek

65
Tiles & Housing Fixtures Manufacturer
INAX Corporation
Tokyo, Japan
1985

AD/Motoo Nakanishi
D/Paos, Inc.

61

62

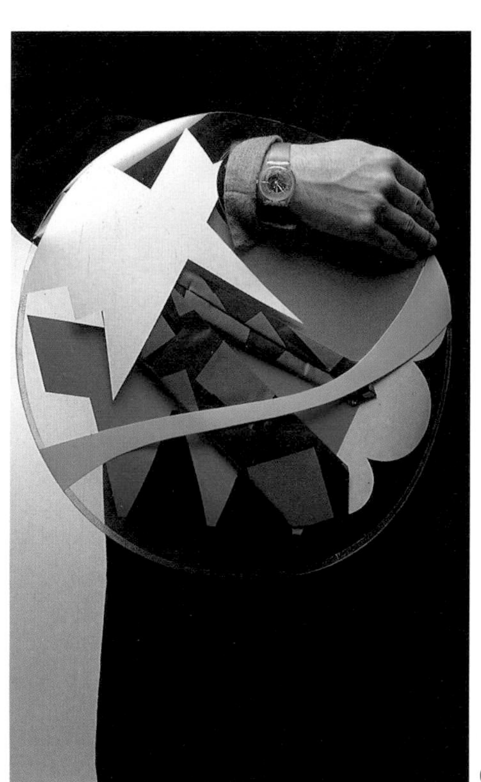
63

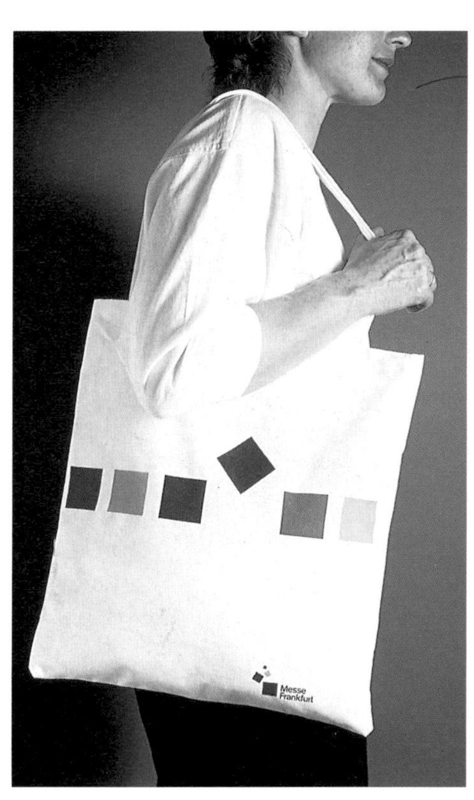

64

65

66・67・68
銀行
ビニール
ナショナル・ウエストミンスター・バンク
イギリス, エセックス州サウスエンド・オン・シー
1987

AD／デーヴィッド・ポックネル
D／ジョナサン・ラッセル
DF／ポックネル・アンド・カンパニー

アクセス, マスターカード, ユーロカードをデザイン要素として使った3点セットのバッグ。

66・67・68
Bank
Plastic
National Westminster Bank
Southend-on-Sea, Essex, England
1987

AD／David Pocknell
D／Jonathan Russell
DF／Pocknell & Co.

This set of three similar bags uses the Access/MasterCard/Eurocard and NatWest logos, along with popular graphic symbols, to highlight areas of convenient use for the Access card and system.

66

67

68

69

69
ブティック"ヘックツ"
ザ・メイ・デパートメント・ストアー
アメリカ, セントルイス
1987

D/ステッフ・ガイスビューラー
DF/シャマイエフ・アンド・ガイスマー・アソシエイツ
この"ヘックツ"のショッピングバッグは，メイ百貨店が所有する大部分のショップのためのデザイン・システムの一環として制作されたもの。ダイヤモンドの模様が入っており，色は白とワインカラー。紙はピーク紙。下部に店のロゴが入っている。

70
ブティック
ポリエチレン
ビッキー
神戸市
1987

D/三浦直美
DF/尾崎紙工所

69
Boutique "Hecht's"
The May Department Store Co.
St. Louis, USA
1987

D/Steff Geissbuhler
DF/Chermayeff & Geismar Associates
The shopping bag for Hecht's is part of an overall design system for most of the stores owned by The May Department Store Co. In each case, the diamond pattern is used with the white and wine-colored palette on a pique paper, while the individual store logos are added at the bottom.

70
Boutique
Plastic
Bikky
Kobe City, Japan
1987

D/Naomi Miura
DF/Ozaki Paper Products Co., Ltd.

70

71
ファッション・ブティック
ザッカラ
オーストラリア，シドニー

D／キース・デーヴィス
DF／デーヴィス・ファーレル・プラス・アソシエイツ

72
ブティック
バニティ
神戸市
1987

DF／尾崎紙工所

71
Fashion Boutique
Zuccala
Sydney, Australia

D／Keith Davis
DF／Davis Farrell+Associates

72
Boutique
Vanity
Kobe City, Japan
1987

DF／Ozaki Paper Products Co., Ltd.

71

72

73

74

73・74
デパート
ブルーミングデイルズ
アメリカ, ニューヨーク
1985

AD／ジョン・ジェイ
D／ティム・ガーヴィン
DF／ティム・ガーヴィン・デザイン

73・74
Department Store
Bloomingdale's
New York, USA
1985

AD／John Jay
D／Tim Girvin
DF／Tim Girvin Design, Inc.

75
デパート
ノードストローム
アメリカ, シアトル
1986

AD／チェリル・フジイ
D／シドニー・ハマーキスト
I／トニー・キンバル
Calligrapher／ブルース・ヘイル
A／ザ・パック・コーポレーション

75
Department Store
Nordstrom
Seattle, USA
1986

AD／Cheryl Fujii
D／Sydney Hammerquist
I／Tony Kimball
Calligrapher／Bruce Hale
A／The Pack Corporation

75

76
懐石レストランのテイクアウト用バッグ
赤坂米穀(グランジュ・ポーター・ジャポネーズ)
東京
1987

AD／伊藤彰英
D／原田健次
DF／ビア・ボーリンク
和食を新しいシステムでオペレーションする店ということで，純粋な和風イメージは避け，機能的なサイズと形で，ロゴを使用したシンプルなデザイン。

77
ブティック
ワールド・リザ・ドルチェ
神戸市
1987

D／ワールド・ドルチェ企画室
DF／尾崎紙工所

76
Japanese Food Service Company
Akasaka Beikoku Co., Ltd.
(Grange Porter Japonais)
Tokyo, Japan
1987

AD／Akihide Ito
D／Kenji Harada
DF／Via Borink Co., Ltd.
A shop which operates a totally new system for serving Japanese food ; avoiding the traditional Japanese image, the symbol has been designed using a logo of functional size and shape.

77
Boutique
World Liza Dorche
Kobe City, Japan
1987

D／World Dorche Planning Room
DF／Ozaki Paper Products Co., Ltd.

76

77

78
ブティック
シェトワ
大阪
1985

DF／ビア・ボーリンク

79
ファッション・ブティック
アンジェラ・グラショフ
西ドイツ，シュトウットガルト
1986

D／ピエール・メンデル
DF／メンデル・アンド・オーベラー

78
Boutique
Chez Toi Co., Ltd.
Osaka, Japan
1985

DF／Via Borink Co., Ltd.

79
Fashion Boutique
Angela Grashoff
Stuttgart, West Germany
1986

D／Pierre Mendell
DF／Mendell & Oberer

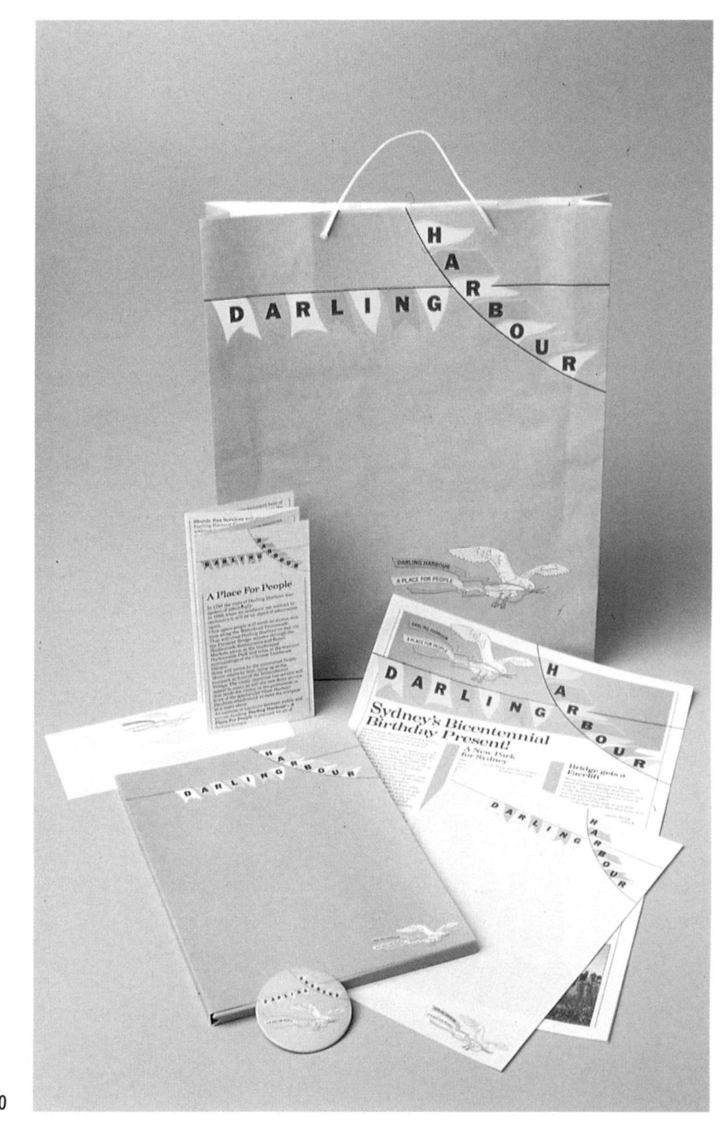

80

81

80
"ダーリン・ハーバー"
ダーリン・ハーバー・オーソリティ
オーストラリア, シドニー

D／マイケル・ファーレル
DF／デーヴィス・ファーレル・プラス・アソシエイツ

81
バラエティ・ショップ
サンリオ(サンリオ・ギャラリー)
東京
1987

D／サンリオ市場開発部

80
"Darling Harbour"
Darling Harbour Authority
Sydney, Australia

D／Michael Farrell
DF／Daris Farrel+Associates

81
Variety Shop
Sanrio Co., Ltd. (Sanrio Gallery)
Tokyo, Japan
1987

D／Sanrio Co., Ltd.

82

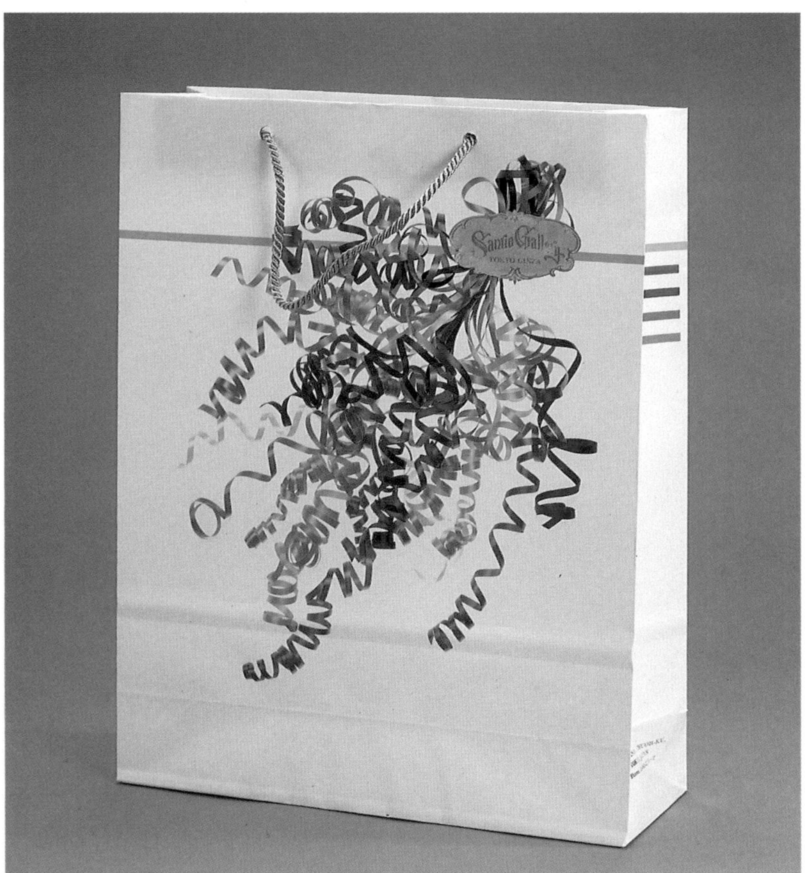
83

82
デパート
ノードストローム
アメリカ, シアトル
1986

AD/チェリル・フジイ
D/ティム・ガーヴィン
DF/ティム・ガーヴィン・デザイン

83
デパートのチョコレート・ブランド用
ショッピングバッグ
ブルーミングデイルズ
アメリカ, ニューヨーク
1981

D/ロバート・ガーシン
DF/ロバート・P・ガーシン・アソシエイツ
ブルーミングデイルズの"オ・ショコラ"のためにつくられたショッピングバッグ。これはブランドのトータル・ラッピング・システムの一環としてデザインされたものであり, システムにはチョコレートを入れるためのボックス類も含まれている。バッグはボックスの銅色とリボンのピンク, 紫, 青緑色に合わせた色になっている。

82
Department Store
Nordstrom
Seattle, USA
1986

AD/Cheryl Fujii
D/Tim Girvin
DF/Tim Girvin Design, Inc.

83
Chocalate Shopping Bag for
Department Store
Bloomingdale's
New York, USA
1981

D/Robert Gersin
DF/Robert P. Gersin Associates, Inc.
The "Au Chocolat" shopping bag was created for Bloomingdale's specialty food department, and was designed to carry the store's hand-packed chocolates. The "Au Chocolat" bag reproduces the copper color of the chocolate box and the pink, purple and turquoise ribbons.

84
ファション・ブランド"ÓKeef"
やまもと寛斎
東京
1987

D／榎本三代子
DF／やまもと寛斎広報宣伝室

85
洋傘袋
ビニール
グレート・ブリテン
東京
1988

D／徳永哲夫
A／フジ紙工

86
ラッピング・グッズ・ショップ
ビニール
ペーパー・ムーン・インターナショナル・ジャパン
東京
1986

D／森下昭彦
DF／ペーパー・ムーン・インターナショナル・ジャパン
メーンの取扱い商品が，カードとラッピング・ペーパー（ロール）であるため，この相反する形状を考慮してデザインしたもの。フランスパンより長いバッグはユニークで，持ってもおしゃれだという考えから，色も派手なブライト・ピンクにしている。

84
Fashion Brand "ÓKeef"
Yamamoto Kansai Co., Ltd.
Tokyo, Japan
1987

D／Miyoko Enomoto
DF／Yamamoto Kansai Co., Ltd.

85
Bag for Umbrella
Plastic
Great Britain Corporation
Tokyo, Japan
1988

D／Tetsuo Tokunaga
A／Fuji Paper Industry Co., Ltd.

86
Wrapping Goods Shop
Plastic
Paper Moon Int'l Japan, Inc.
Tokyo, Japan
1986

D／Akihiro Morishita
DF／Paper Moon Int'l Japan Inc.
Designed for a retailer of wrapping paper and greeting cards, the bags reflect their contents ; they are as long as baggettes and colored a silky, bright pink.

84

85

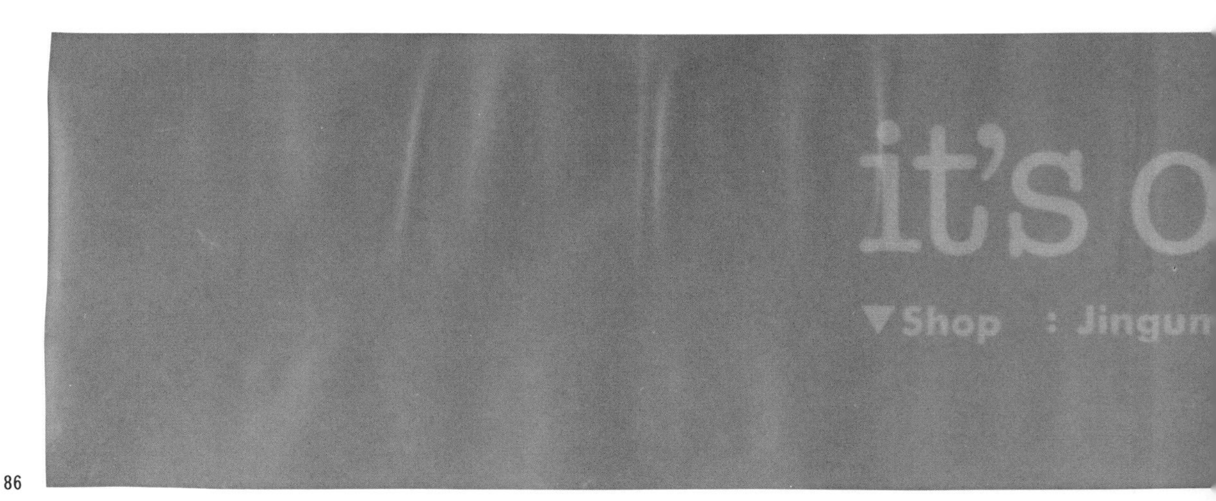

86

87

88

87
アパレル・ブランド
ア・ストア・ロボット(ビリー)
東京
1988

AD/北村一郎
D/占部宏二
DF/ア・ストア・ロボット東京事務所

87
Apparel Brand
A Store Robot Co., Ltd. (Billy)
Tokyo, Japan
1988

AD/Ichiro Kitamura
D/Koji Urabe
DF/Tokyo Office, A Store Robot Co., Ltd.

88
ファッション・ブランド"ロペ"
ジュン
東京
1986

AD/井上嗣也
DF/ビーンズ

88
Fashion Brand "Ropé"
Jun Co., Ltd.
Tokyo, Japan
1986

AD/Tsuguya Inoue
DF/Beans Co., Ltd.

AT-BRITAIN

nly paper moon...
4-26-35 FARM/3F, Shibuya-ku, Tokyo:150 Phone(03)478-4238

89

90

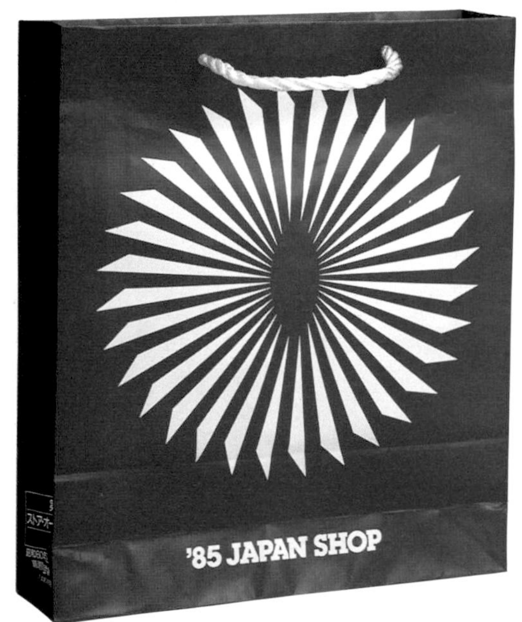

89
市販用ショッピングバッグ
松城紙袋工業
東大阪市
1986

I／森 貞人

89
Shopping Bags
Matsushiro Paper Bag Industry Co., Ltd.
Osaka, Japan
1986

I／Sadahito Mori

90
店舗総合見本市"ジャパン・ショップ"
日本経済新聞社
東京
1983—1985

D／永井一正

90
Comprehensive Store Exhibition "Japan Shop"
Nihon Keizai Shimbun, Inc.
Tokyo, Japan
1983—1985

D／Kazumasa Nagai

91
デパートのバレンタイン・デー用バッグ
三越
東京
1987

AD／祖父江将浩
I／舟橋全二
DF／博報堂
デパートのバレンタイン・デーのためのショッピングバッグで親しみのある素材とハート型を組み合わせ，明るい印象になるようデザインされた。

92
デパートのオープニング・キャンペーン
西武百貨店（所沢西武）
埼玉県所沢市
1986

D／松永 真
Copywriter／日暮真三
デパート"所沢西武"のオープニングキャンペーン用のショッピングバッグ。都心への通勤者が多い客層に合わせて閉店時間を遅くすることをシンボル化することがコンセプトの主眼とされた。

91
Department Store Valentine's Day Shopping Bag
Mitsukoshi Limited
Tokyo, Japan
1987

AD／Masahiro Sobue
I／Zenji Funabashi
DF／Hakuhodo, Inc.
A Valentine's Day shopping bag for a department store. The design creates an impression of cheerfulness by using appealing materials and heart motif.

92
Department Store Opening Campaign
The Seibu Department Stores Co., Ltd. (Tokorozawa Seibu)
Tokorozawa City, Japan
1986

D／Shin Matsumaga
Copywriter／Shinzo Higurashi
A shopping bag for the opening campaign of the department store "Tokorozawa Seibu". Considering the fact that most of the population of the city are commuters to downtown Tokyo, the store has set the closing time at the late evening. Hence, the design concept is placed to symbolize the characteristic of the store concept about the closing time.

91

92

昼と夜、あなたは二つの顔を持つ。

昼と夜、あなたは二つの顔を持つ。

所沢西武4月5日㊎オープン。

93

94

95

93・94
ブティック
ジュン(グレタ・ガルボ)
東京
1987

AD/戸田正寿
D/水島正則, 樋口真貴
DF/戸田事務所, リネアフレスカ

95
照明会社
布
プリモ・ライティング
アメリカ, コネチカット州ノーウォーク
1986

D/フレッド・トローラー
DF/フレッド・トローラー・アソシエイツ

93・94
Boutique
Jun Co., Ltd.(Greta Galubot)
Tokyo, Japan
1987

AD/Masatoshi Toda
D/Masanori Mizushima, Masataka Higuchi
DF/Toda Studio, Linea Fresca

95
Lighting Company
Cloth
Primo Lighting
Norwalk, CT, USA
1986

D/Fred Troller
DF/Fred Troller Associates

96

97

98
ギフト・ショップ
ポリエチレン
ハニー(キネティクス)
東京
1982

AD／佐久間由紀子
D／滝 由佳理, 友枝康二郎
DF／インタレスト

98
Gift Shop
Plastic
Honey Co., Ltd. (Kinetics)
Tokyo, Japan
1982

AD／Yukiko Sakuma
D／Yukari Taki, Kojiro Tomoeda
DF／Interest Inc.

96
ファション・ブランド "Kansai Yamamoto"
やまもと寛斎
東京
1987

D／榎本三代子
DF／やまもと寛斎広報宣伝室

97
雑貨店
ポリエチレン
ハニー(T.P.O.)
東京
1987

AD／佐久間由紀子
D／滝 由佳理
DF／インタレスト

96
Fashion Brand "Kansai Yamamoto"
Yamamoto Kansai Co., Ltd.
Tokyo, Japan
1987

D／Miyoko Enomoto
DF／Yamamoto Kansai Co., Ltd.

97
Miscellaneous Goods Shop
Plastic
Honey Co., Ltd. (T.P.O.)
Tokyo, Japan
1987

AD／Yukiko Sakuma
D／Yukari Taki
DF／Interest Inc.

98

99
ファッション・ブランド
イッセイミヤケ・アンド・アソシエーツ
東京
1974

D／三宅一生
Typographer／金井 淳

100
テナントビルのリフレッシュ・オープン・キャンペーン
岡田屋モアーズ
横浜・川崎
1985

AD／杉本英介
D／永井裕明
DF／ブレックファースト

101
ジーンズ・ショップ
ビニール
ジョイント
大阪

AD／住田芳明
DF／ジョイント

102
ブティック
ソニー・プラザ(James Stride)
東京
1987

D／渡辺通正
男性のためのコンセプト・ショップ。店名の"James Stride"は，自分の歩調でゆっくり歩く，自分の生活スタイルをもった男性をイメージしている。ブルーは"男性"を，ゴールドは"高級感"を，全体の大理石模様は"永遠に変わらないもの"を表現している。

103
子供服店
ビニール
リトルアンデルセン(ヒステリック・ミニ)
東京
1988

AD／木島 努
D／富田勝利，佐々木克利
I／木島 努
DF／リトルアンデルセン・ヒステリック・ミニ事業部

104
ファッション・ブランド"フランドル・クラブ 1978"
ビニール
フランドル
東京
1978

D／フランドル企画室

99
Fashion Brand
Issey Miyake & Associates Inc.
Tokyo, Japan
1974

D／Issey Miyake
Typographer／Kiyoshi Kanai

100
Boutique Shopping House Re-Opening Campaign
Okadaya-More's
Yokohama, Japan
1985

AD／Eisuke Sugimoto
D／Hiroaki Nagai
DF／Breakfast Co., Ltd.

101
Jeans Shop
Plastic
Joint Co., Ltd.
Osaka, Japan

AD／Yoshiaki Sumida
DF／Joint Co., Ltd.

102
Boutique
Sony Plaza Co., Ltd. (James Stride)
Tokyo, Japan
1987

D／Michimasa Watanabe
A conept shop for men. The shop's name, "James Stride", has as its image men who like to lead their individual lifestyles at their own pace. Blue expresses "Masculinity", gold suggests "High Quality" while the marble pattern creates an impression of goods of everlasting quality.

103
Children's Clothing Shop
Plastic
Little Andersen Co., Ltd. (Hysteric Mini)
Tokyo, Japan
1988

AD／Tsutomu Kijima
D／Katsutoshi Tomita, Katsutoshi Sasaki
I／Tsutomu Kijima
DF／Hysteric Mini Dept., Little Andersen Co., Ltd.

104
Fashion Brand "Flandre Club 1978"
Plastic
Flandre Co., Ltd.
Tokyo, Japan
1978

D／Planning Room, Flandre Co., Ltd.

99

100

101

102

103

104

105
ブティック
ビニール
ジーピー(GP-8渋谷店)
埼玉県川口市
1987

D/呉 智子
DF/近宣
A/白石商店

106
婦人靴店
アジアの靴
名古屋
1973

D/立松 脩
DF/デコム・クリエーティブアート&サービス

107
駄菓子屋
ハニー(ハラッパA)
東京
1984

AD/佐久間由紀子
D/友枝康二郎
I/鴨沢祐仁
DF/インタレスト
駄菓子屋と昔なつかしいおもちゃを扱っている店。バッグにあしらった"クシー君"は,ファンタジックで,ノスタルジックな独特の世界をもつ"ハラッパA"のイメージキャラクター。

105

106

107

108
Tシャツ・ショップ
明石企画
名古屋
1973

D／小島 歩
DF／まるまる図案

109
スウェットウェア専門店
ゼロ・ワン
東京
1987

AD／荒木隆一
D／石井久美子
I／北田哲也
DF／イフコ・デザインハウス
A／ヘルム・エージェンシー
スウェットウェア専門店が10周年を迎え，その喜びと感謝の気持を表現するために，音譜や，いろいろにデフォルメされたキャラクターを配した。

108

105
Boutique
Plastic
G & P Co., Ltd. (GP-8 Shibuya)
Kawaguchi City, Japan
1987

D／Tomoko Kure
DF／Kinsen Co., Ltd.
A／Shiraishi Shoten Co., Ltd.

106
Ladies' Shoes Store
Asian Shoes, Co., Ltd.
Nagoya, Japan
1973

D／Osamu Tatematsu
DF／Decom Creative Art & Service Co., Ltd.

107
Sweets and Toys Shop
Honey Co., Ltd. (Harrapa A)
Tokyo, Japan
1984

AD／Yukiko Sakuma
D／Kojiro Tomoeda
I／Yuji Kamosawa
DF／Interest Inc.
This bag is for a shop selling sweets and toys from yesteryear. Mr. Xie is the image character of Harrapa A, where a world of Fantasia and Nostalgia can be found.

108
T-Shirt Shop
Akashi Kikaku Co., Ltd.
Nagoya, Japan
1973

D／Ayumu Kojima
DF／Maru Maru Zuan

109
Sweat Wear Shop
Zero One Co., Ltd.
Tokyo, Japan
1987

AD／Ryuichi Araki
D／Kumiko Ishii
I／Tetsuya Kitada
DF／Ifco Design House Co., Ltd.
A／Helm Agency
This shop dealing exclusively in Sweat Wear is celebrating its 10th anniversary ; so, in order to express their joy and gratitude, the bag has been designed with musical manuscripts and various deformation.

109

110

110
ファッション・ショップ
バーニーズ・ニューヨーク
アメリカ, ニューヨーク
1983

D／トム・ガイスマー
DF／シャマイエフ・アンド・ガイスマー・アソシエイツ
ファッション・ブティック"バーニーズ・ニューヨーク"のC.I. プログラムの一環として制作されたショッピングバッグ。

110
Fashion Retailer
Barney's New York
New York, USA
1983

D／Tom Geismar
DF／Chermayeff & Geismar Associates
The shopping bags for Barney's New York were part of an overall corporate identity program done for this clothing store.

111
京都"ルネッサンス・プラザ"
林原グループ
岡山市

AD／リョウ・ウラノ
DF／ウラノ・コミュニケーション・インターナショナル
古都京都に文化活動のための新しい環境をつくることを目的としたプロジェクト"ルネッサンス・プラザ"のためにデザインされたもの。三角形のロゴタイプは建築物がベースになっている。

111
"Renaissance Plaza" in Kyoto
Hayashibara Group
Okayama City, Japan

AD／Ryo Urano
DF／Urano Communication Internatinal
This project involved naming a plaza designed to create a new environment for cultural activities in Kyoto, the old Capital city of Japan. The triangle logotype was derived from the unique architectural feature.

111

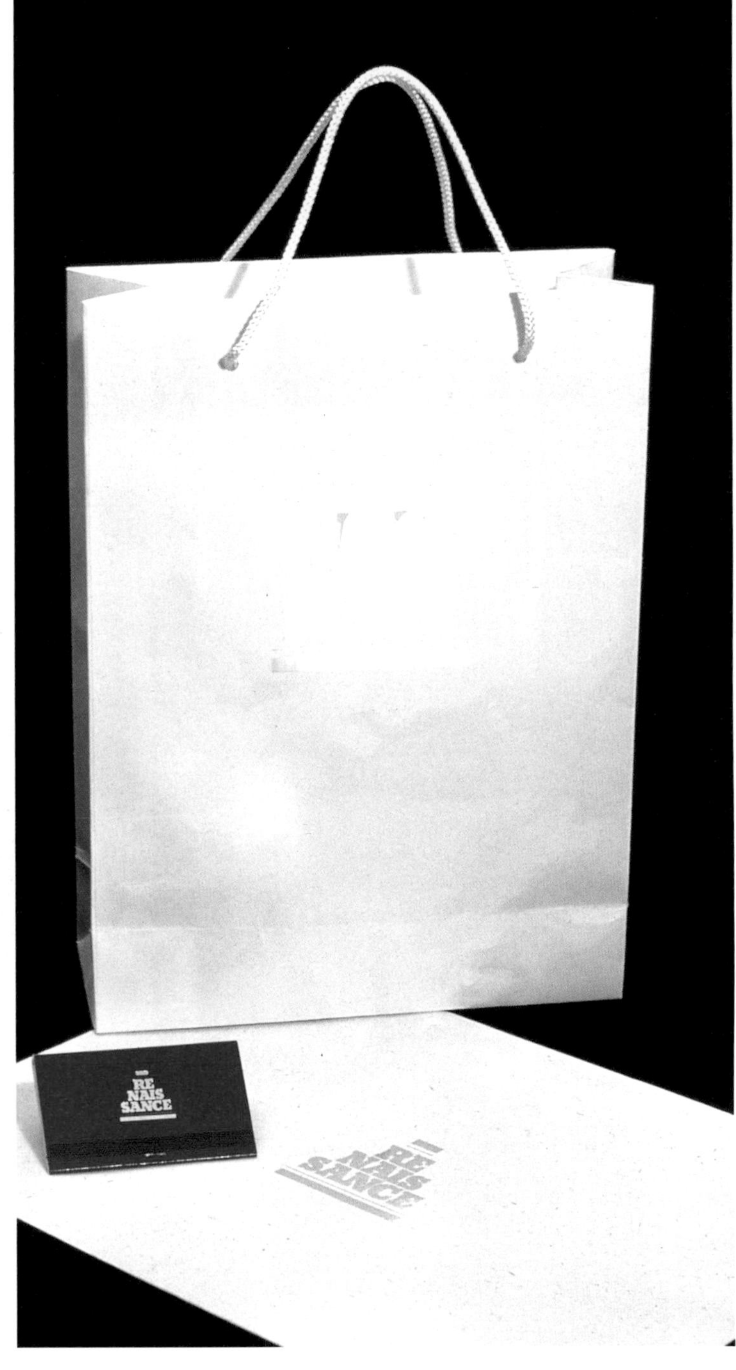

112

112
市販用"Free Time"
ビニール
大和シュガル
東京
1984

D／アペックス

113
洋菓子店
ビアンクール
神戸市
1986

AD／宇田 優
D／奥田一明
DF／ユウ・デザイン事務所

112
Shopping Bag "Free Time"
Plastic
Sugal Creative Production
Tokyo, Japan
1984

D／Apex Co., Ltd.

113
Confectionery
Billancourt Limited
Kobe City, Japan
1986

AD／Masaru Uda
D／Kazuaki Okuda
DF／U Design Office

113

114
南米民芸品店
チチカカ
東京
1980

D／関本ユキエ
DF／チチカカ
ペルーのチャンカイ文化(インカ時代以前，AD1100〜1450年頃)の伝統的な図柄。この時代に見られるさまざまなユニークなモチーフのうちのひとつをシンボルにしている。

114
South American Folk Craft Shop
Titicaca
Tokyo, Japan
1980

D／Yukie Sekimoto
DF／Titicaca
Traditional patterns from the Chankai culture of Peru (before the Inca Period AD 1100 to about 1450). This is one of the various symbols to be found among th unique motives of this era.

114

115

115
モータースポーツ・グッズ・ショップ
原宿サーキット
東京
1987

D／米澤公二
DF／ホット・スタッフ
A／モータースポーツ・システム

115
Motor Sporting Goods Shop
Harajuku Circuit
Tokyo, Japan
1987

D／Koji Yonezawa
DF／Hot Staff
A／Motor Sports System

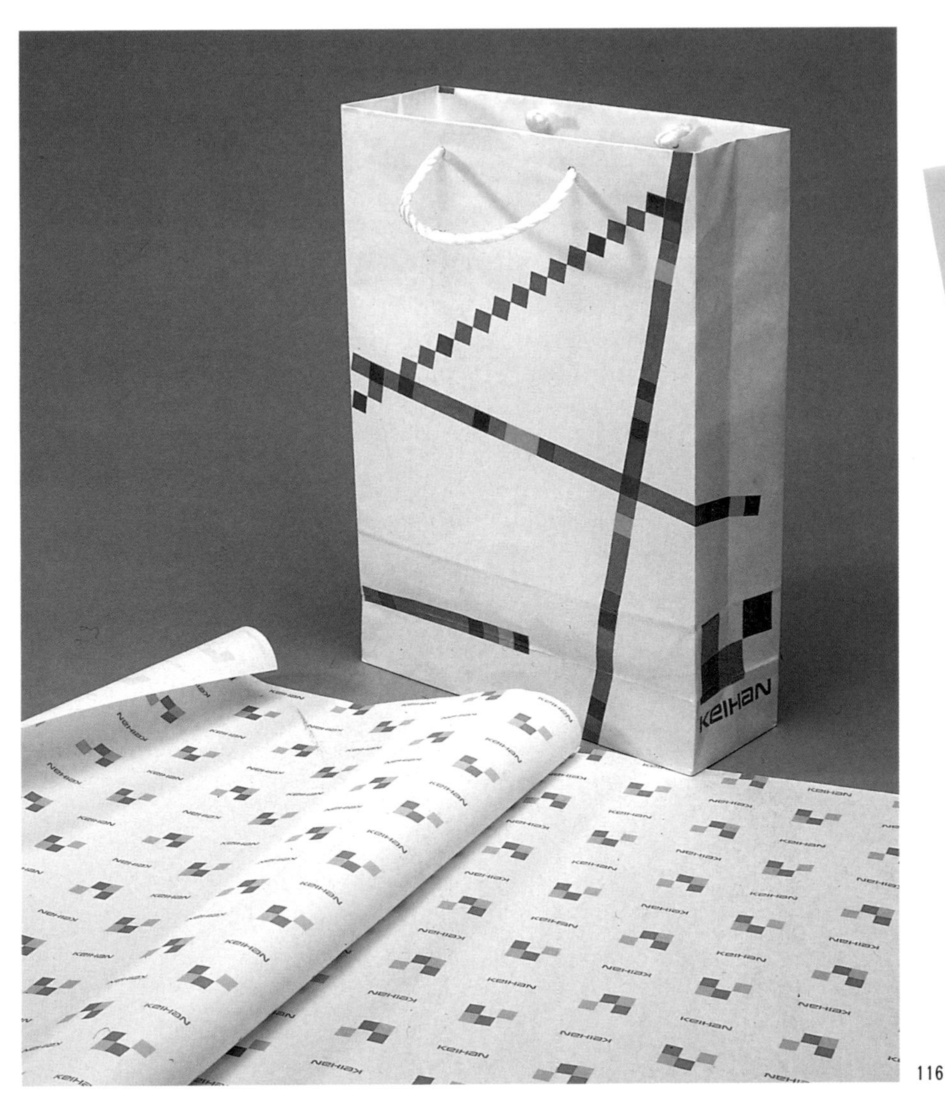

116

117

118

116・118
デパート
京阪百貨店
大阪
1985

D／松永 真
駅の操車場跡に街づくりと平行しながらつくられた京阪百貨店のCI計画のアプリケーションとしてのショッピングバッグと包装紙。"スクウェア"がデザインの基本モチーフとして表現されている。レギュラータイプと，販売用のものがある。

117
ブティック
ビニール
プロライン(元町ポーター・クラブ)
横浜
1987

D／プロライン

116・118
Department Store
Keihan Department Stores Ltd.
Osaka, Japan
1985

D／Shin Matsunaga
This shopping bag and wrapping paper are based on the same concept as the C.I. plan for the Keihan Department stores, which were built as part of a redevelopment project for the site of the former station yard. There are regular types and the types for sale.

117
Boutique
Plastic
Pro-Line Inc. (Motomachi Poter Club)
Yokohama, Japan
1987

D／Pro-Line Inc.

119

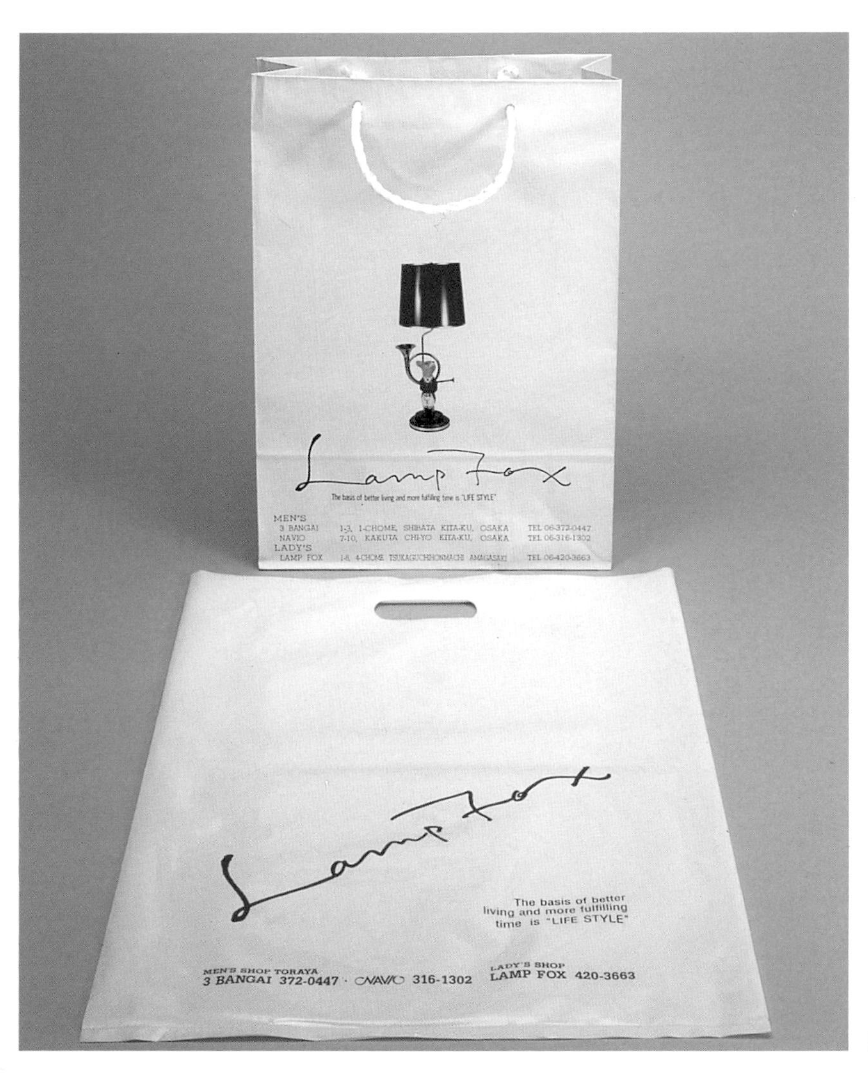

120

119
靴店
ビニール
フロンティア
東京
1986

D／真水真樹
DF／東京工房

120
メンズ・ブティック
紙，ビニール
メンズショップ・トラヤ(Lamp Fox)
尼崎市
1983(紙袋)
1987(ビニール袋)

D／Lamp Fox

119
Shoe Store
Plastic
Frontier
Tokyo, Japan
1986

D／Maki Masui
DF／Tokyo Kobo Ltd.

120
Men's Boutique
Paper, Plastic
Men's Shop Toraya Co., Ltd.
(Lamp Fox)
Amagasaki City, Japan
1983 (Paper Bag)
1987 (Plastic Bag)

D／Lamp Fox

121

122

121
キャラクター・グッズ・ショップ
ビニール
クラブ・ジャップス
神戸市
1979

D／クラブ・ジャップス

122
ブティック
ビニール，紙
Wooden Doll
東京
1987

D／鈴木 淳
Tシャツのプリント・デザインを企画中に生まれたショッピングバッグ・デザインで，店のマークにも使用されるようになった。

121
Character Goods Shop
Plastic
Club Japs Inc.
Kobe City, Japan
1979

D／Club Japs Inc.

122
Boutique
Plastic, Paper
Wooden Doll
Tokyo, Japan
1987

D／Atsushi Suzuki
This shopping bag design came from sketches for T-shirt prints which have now been adopted as the shop's logo.

123

123
バラエティ・ショップ
サンリオ（ギフト・ゲート）
東京
1987

D／サンリオ営業推進部

124
バラエティ・ショップ
ビニール
ウィークエンズ
東京
1985

D／近藤絹衣
DF／ウィークエンズ
A／エリス

123
Variety Shop
Sanrio Co., Ltd. (Gift Gate)
Tokyo, Japan
1987

D／Sanrio Co., Ltd.

124
Variety Shop
Plastic
Weekends
Tokyo, Japan
1985

D／Kinue Kondo
DF／Weekends
A／Elice Co., Ltd.

124

125

126

127

128

125—128
ファッション会社
ビニール
エスプリ・デ・コーポレーション
アメリカ，サンフランシスコ
1984—1987

D／タモツ・ヤギ
DF／エスプリ・グラフィック・デザイン・ステュディオ

129
デパート
ブルーミングデイルズ
アメリカ，ニューヨーク
1972

D／マッシモ・ヴィネリ
DF／ヴィネリ・アソシエイツ

130
小売店
プランジーズ
アメリカ，ウィスコンシン州マジソン

D／ケネス・ラブ
DF／アンスパック・グロスマン・ポーチュガル・インコーポレイテッド
店名の力強いタイポグラフィーが真っ先に目に飛び込んでくるようにデザインされている。メーンカラーはナチュラルグレーで，多色刷のギフト・ボックスの色あいを引きたたせる役目をもっている。

125—128
Fashion Company
Plastic
Esprit De Corporation
San Francisco, USA
1984—1987

D／Tamotsu Yagi
DF／Esprit Graphic Design Studio

129
Department Store
Bloomingdale's
New York, USA
1972

D／Massimo Vignelli
DF／Vignelli Associates

130
"Prange's" Store
Prange's
Madison, Wisu., USA

D／Kenneth Love
DF／Anspach Grossman Portugal Inc.
A large and dominant typographic treatment fo the store's name is the primary visual element used for Prange's packaging. The shopping bag is neutral gray and complements the multi-color treatment used on gift boxes.

129

130

131

132

131
デパートの特別招待会用バッグ
岩田屋
福岡市
1983—1988

D／稲葉欣司
DF／岩田屋販促部
A／ザ・パック

132
ブティック
ベーリー・ストックマン
東京
1986

AD／井上義朗
D／草野和芳
DF／ル・ギャマン

131
Bag for a Department Store Special Invitation Day
Iwataya Department Store Co., Ltd.
Fukuoka City, Japan
1983—1988

D／Kinji Inaba
DF／Iwataya Department Store Co., Ltd.
A／The Pack Corporation

132
Boutique
Bailey Stockman Co., Ltd.
Tokyo, Japan
1986

AD／Yoshiro Inoue
D／Kazuyoshi Kusano
DF／Le Gamin

133

133
チョコレート・ショップ
ショコラティエ・マサール
札幌市

D／谷津松枝
DF／イメージ・イン・スタジオ

133
Chocolate Shop
Chocolatier Masále
Sapporo City, Japan

D／Matsue Yatsu
DF／Image Inn Studio Co., Ltd.

134
ファッションビルのオープニング・キャンペーン
紙，布
心斎橋 Vivre 21
大阪
1987

AD／永井ヒロ子
D／本多厚士
DF／森久第二事務所
オープン時にプレミアムバッグとして使用。赤と緑（内面），ロゴ，チロリアンリボンで，派手さ，楽しさを表現。厚手の紙を使用。

134
Boutique Shopping House
Opening Campaign
Paper, Cloth
Shinsaibashi Vivre 21
Osaka, Japan
1987

AD／Hiroko Nagai
D／Atsushi Honda
DF／Morihisa Second Co., Ltd.
This is used as a premiun bag for a store-opening sale. The colors are red (outside) and green (inside) ; the logo and the Tyrolean ribbon give a kind of showy, dressy and pleasant mood.

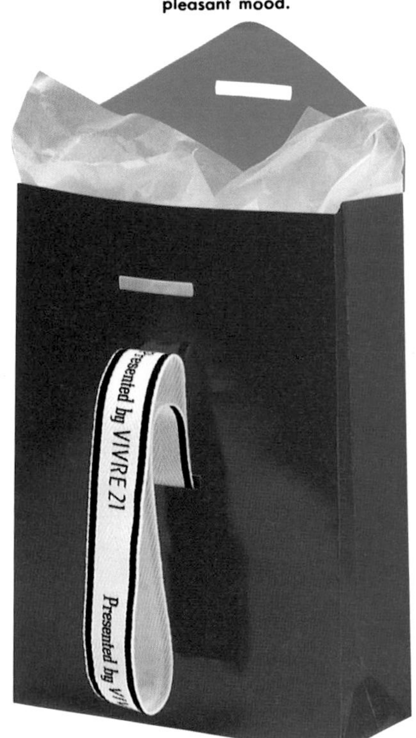

134

135

136

135
展示会用のショッピングバッグ
スーパーバッグ
東京
1987

AD/スーパーバッグS.P.C.室
D/石崎洋之
DF/ヴァーナル・アンド・カンパニー

136
ファッション販売会社
バークレイ
東京
1987

Producer/工藤高子
D/古田 修
都市部を中心に10数店舗を展開するファッション販売会社のショッピングバッグで，ショップ・アイデンティティの確立とCI計画の一環性のなかで，デザインされた。商品群と客層から想定されたベージュを基本カラーに選び，メインカラーに白を使ったシンボルロゴタイプがあしらわれている。

135
Shopping Bag for an Exhibition
Super Bag Co., Ltd.
Tokyo, Japan
1987

AD/S.P.C.Room, Super Bag Co., Ltd.
D/Hiroyuki Ishizaki
DF/Vernal And Company Ltd.

136
Fashion Dealer
Berkeley Co., Ltd.
Tokyo, Japan
1987

Producer/Takako Kudo
D/Osamu Furuta
A shopping bag designed as one ked item to estabilsh the identity of the shop and promoting the C. I. project of Berkeley Co., Ltd. who operate more than ten shops, mainly in urban areas. The symbol/logotype, mainly in white, is coordinated with beige, which was determined as the base color after taking consideration the nature of the products and the type of customers.

137
ステーショナリー・ショップ
紙, ビニール
パルコ
東京
1986

D／塚本佳哉樹
DF／コミュニケーションアーツR
ショップ名である"犀"の字のもつイメージをもとにデザインされたショッピングバッグと包装紙。

138
ブティック
ベーリー・ストックマン
東京
1977

AD／井上義朗
D／草野和芳
DF／ル・ギャマン

137
Stationery Shop
Paper, Plastic
Parco, Inc.
Tokyo, Japan
1986

D／Kayaki Tsukamoto
DF／Communication Arts "R"
Shopping bag and wrapping paper using the name of the shop written with a Chinese character as the basic motif.

138
Boutique
Bailey Stockman Co., Ltd.
Tokyo, Japan
1977

AD／Yoshiro Inoue
D／Kazuyoshi Kusano
DF／Le Gamin

137

138

139

140

139
美術館
グラフィーシェ・ザムルンク・アルバーティナ
オーストリア，ウィーン
1985

D／ハリー・メッツラー
DF／ハリー・メッツラー・アートデザイン
ウィーンの"アルバーティナ"はクラシカルなグラフィック・アート作品を収蔵する世界的に有名な美術館。これは，ある展示会のためにつくられたバッグで，"A"はアルバーティナのイニシャルを示している。

140
衣料品会社
サンフランシスコ・クロージング
アメリカ，ニューヨーク
1987

D／ジョージ・ジャーニー
DF／ジョージ・ジャーニー・インコーポレイテッド

141
ファッションビル
大宮ステーションビル
埼玉県大宮市
1986

D／八木健夫
DF／オフィース・ピーアンドシー

141

142
"IDCNY"
インターナショナル・デザイン・センター/ニューヨーク
アメリカ, ニューヨーク
1985

AD/マッシモ・ヴィネリ, マイケル・ビエルト
Cordinator/ファーン・マリス
DF/ヴィネリ・アソシエイツ

143
アパレル・ブティック
ビニール
Voxel Yamamoto
京都

D/ワールド宣伝部

139
Museum
Graphische Sammlung
Albertina
Vienna, Austria
1985

D/Harry Metzler
DF/Harry Metzler Artdesign
The "Albertina" is one of the world most famous museums for classical graphic arts. The bag was produced for the occasion of an exhibition. 'A' stands for Albertina.

140
Clothing Company
San Francisco Clothing
New York, USA
1987

D/George Tscherny
DF/George Tscherny, Inc.

141
Shopping Center
Omiya Station Building Inc.
Omiya City, Japan
1986

D/Takeo Yagi
DF/Office P&C Inc.

142
"IDCNY"
Internatinal Design Center/New York
New York, USA
1985

AD/Massimo Vignelli, Michael Bierut
DF/Vignelli Associates

143
Apparel Boutique
Plastic
Voxel Yamamoto
Kyoto, Japan

D/Advertising Dept., World Co., Ltd.

142

143

144

145

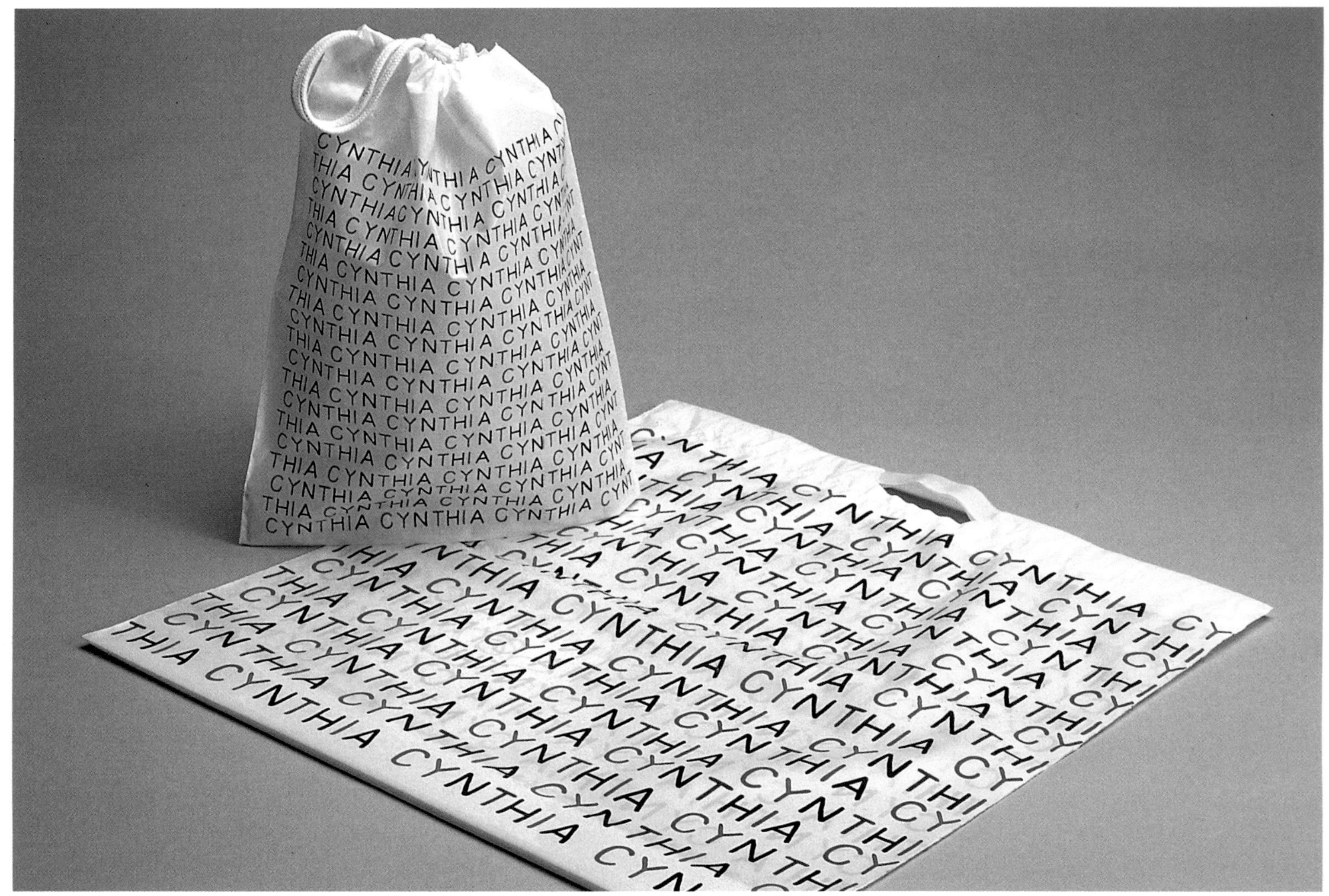

146

144
バラエティ・ショップ
リチャード
大阪
1982

D／横尾忠則

145
"テカデュー"
テカデュー
イタリア，ノバラ
1987

D／ウォルター・バルマー
DF／ウニデザイン

146
ブティック
ビニール
シンシア
東京
1982

D／シンシア企画部

147
アパレル・ブティック
ビニール
ロス
神戸市
1985

DF／Boy Company

148
ブティック
ビニール
ロック座
東京
1986

D／小林敏明
DF／シンコー・ミュージック

149
ブティック
ビニール
ジーピー(パジャブ渋谷店)
埼玉県川口市
1987

D／呉 智子
DF／近宣
A／白石商店

144
Variety Shop
Richard Co., Ltd.
Osaka, Japan
1982

D／Tadanori Yokoo

145
"Tecadue"
Tecadue
Novara, Italy
1987

D／Walter Ballmer
DF／Unidesign

146
Boutique
Plastic
Cynthia Corp. Ltd.
Tokyo, Japan
1982

D／Planning Dept., Cynthia Corp. Ltd.

147
Apparel Boutique
Plastic
Los
Kobe City, Japan
1985

DF／Boy Company

148
Boutique
Plastic
Rock-za Co., Ltd.
Tokyo, Japan
1986

D／Toshiaki Kobayashi
DF／Shinko Music Publishing Co., Ltd.

149
Boutique
Plastic
G & P Co., Ltd. (Pajaboo Shibuya)
Kawaguchi City, Japan
1987

D／Tomoko Kure
DF／Kinsen Co., Ltd.
A／Shiraishi Shoten Co., Ltd.

147

148

149

150

151

150
アパレル・ブランド"ヒステリック・グラマー"
ビニール
オゾン・コミュニティー
東京
1984

D／北村信彦

151
ブティック
45RPM-Studio Co.
東京
1986

D／益山航士
材質の粗悪な感じを生かし，ゴミ袋的な使用を空想してデザインされた逆説的な作品。

150
Apparel Brand "Hysteric Glamour"
Plastic
Ozone Community Co., Ltd.
Tokyo, Japan
1984

D／Nobuhiko Kitamura

151
Boutique
45RPM-Studio Co.
Tokyo, Japan
1986

D／Hiroshi Masuyama
A somewhat paradoxical design, giving the impression of a bin bag in coarse quality materials.

152
書店
ビニール
ロード・アイランド・スクール・オブ・デザイン
アメリカ，ロード・アイランド州プロヴィデンス
1987

D／マルコム・グレア・デザイナーズ・インコーポレイテッド
ロード・アイランド・スクール・オブ・デザインの書店のためにつくられたイメージをもとにデザインされた。円内に文字があるデザインは，屋外サインや横断膜，ステーショナリーなどにも使われている。

153
アンダーウェアショップ
ホームズ・アンダーウェア
東京

D／Deconatic

154
アパレル・ブティック
ビニール
Alicia（ページボーイ）
東京
1987

D／井賀絵里
DF／Alicia・ページボーイ事業部

152
Bookstore
Plastic
Rhode Island School of Design
Providence, RI, USA
1987

D／Malcolm Grear Designers, Inc.
The design incorporates the image developed for the RI School of Design Bookstore. The letters within the circles are used as exterior signs, banners and stationery.

153
Underclothing Shop
Plastic
Homes' Underwears
Tokyo, Japan

D／Deconatic

154
Apparel Boutique
Plastic
Alicia Co., Ltd. (Page Boy)
Tokyo, Japan
1987

D／Eri Iga
DF／Page Boy Operation Division, Alicia Co., Ltd.

152

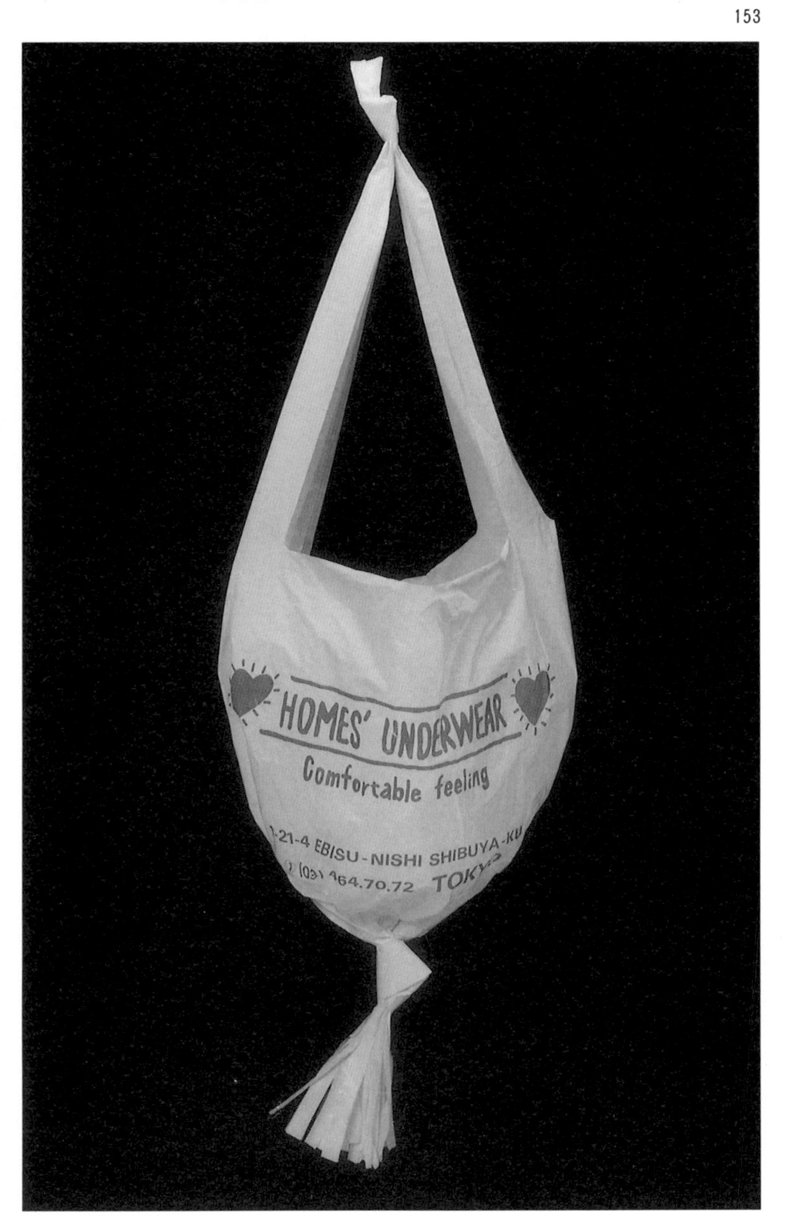

153

154

155
ファション・ブランド"ハイ"
ワコール
東京
1981

D／成瀬始子

155
Fashion Brand "Hai"
Wacoal Corporation
Tokyo, Japan
1981

D／Motoko Naruse

155

156
メンズ・ブティック
ペイス
ホンコン
1983

AD／アラン・チャン
D／アラン・チャン, ベンジャミン・ラウ
DF／アラン・チャン・デザイン
"ペイス"はヨーロッパの衣服を輸入, 販売しているメンズ・ブティック。ショップのインテリアに合わせて, 花こう岩の模様がバッグに使われている。2色刷りで印刷されたこの作品は, 4色刷りに迫る効果をねらって, 数種のスクリーントーンが多用されている。

156
Men's Boutique
Pace
Hong Kong
1983

AD／Alan Chan
D／Alan Chan, Benjamin Lau
DF／Alan Chan Design Co.
"Pace" is a men's boutique selling imported clothes from Europe. The granite effect on the bag is consistent with the shop's interior design. The bag is printed in two colors, gray and black ; use of various screens creates a four-color printing effect.

156

157
ファッション・ブティック
ハリーズ
西ドイツ, ミュンヘン
1985

D／ピエール・メンデル
DF／メンデル・アンド・オーベラー

157
Fashion Boutique
Harry's
Munich, West Germany
1985

D／Pierre Mendell
DF／Mendell & Oberer

158
化粧品会社'84秋のキャンペーン
コーセー化粧品
東京
1984

D／今村倶子
P／中島敏夫
DF／コーセー化粧品

158
Cosmetics Company '84
Autumn Campaign
Kose Cosmetics Co., Ltd.
Tokyo, Japan
1984

D／Tomoko Imamura
P／Toshio Nakajima
DF／Kose Cosmetics Co., Ltd.

157

158

159
ワールド・エクスポ'88
オーストラリア郵政省
オーストラリア, ブリスベーン
1988

D／エメリー・ビンセント・アソシエイツ
P／ジェームズ・キャント
DF／エメリー・ビンセント・アソシエイツ
"オーストラリア・ポスト"ということばが、ビジュアル・アイデンティティの主要なエレメントになっており，テーマや内容は二義的な扱いになっている。このデザインは，郵政活動とテクノロジーの相互関係をシンボル化したもの。

159
Australia Post At World Expo '88
Australia Post
Brisbane, Australia
1988

D／Emery Vincent Assoc.
P／James Cant
DF／Emery Vincent Assoc.
The words Australia Post are the dominant element of this visual identity, with the thematic content incorporated in secondary fashion. The graphic symbolism relates to the activities of postage and technology, concurrently creating feeling of festivity and high spirits appropriate to the ambience of this dynamic occasion.

159

160

160
酒販売店用のショッピングバッグと包装紙
ニッカウヰスキー
東京
1987

AD／岡田宏三
D／岡田宏三, 西原裕彰
DF／オーディ

160
Shopping Bag & Wrapping Paper
The Nikka Whisky Distilling, Co., Ltd.
Tokyo, Japan
1987

AD／Kohzo Okada
D／Kohzo Okada, Hiroaki Nishihara
DF／OD Incorporated

161

161
"マークトガッセ・デュベンドルフ"
ビニール
マークトガッセ・デュベンドルフ
スイス, チューリッヒ
1984

AD／ゴットショーク+アッシュ・インターナショナル
D／フリッツ・ゴットショーク, ハンス・グリュニンガー
DF／ゴットショーク+アッシュ・インターナショナル

162
カジュアルウェア・ショップ
パイ・ブティック
ホンコン
1985

AD／アラン・チャン
D／アラン・チャン, アルヴィン・チャン
DF／アラン・チャン・デザイン
"パイ・ブティック"は, ホンコン製カジュアルウェアのショップ。バッグのデザインに見られる連続的なパターンは, カジュアルウェアの不変性を象徴したもの。

161
"Marktgasse Dübendorf"
Plastic
Marktgasse Dübendorf
Zürich, Switzerland
1984

AD／Gottschalk+Ash Int'l
D／Fritz Gottschalk, Hans Grüninger
DF／Gottshalk+Ash Int'l

162
Casual Wear Shop
Pye Boutique
Hong Kong
1985

AD／Alan Chan
D／Alan Chan, Alvin Chan
DF／Alan Chan Design Co.
Pye Boutique carries Hong Kong-made casual wear for men and women. The pattern repeated on the surface of the shopping bag symbolizes the never-ending energy of this casual wear.

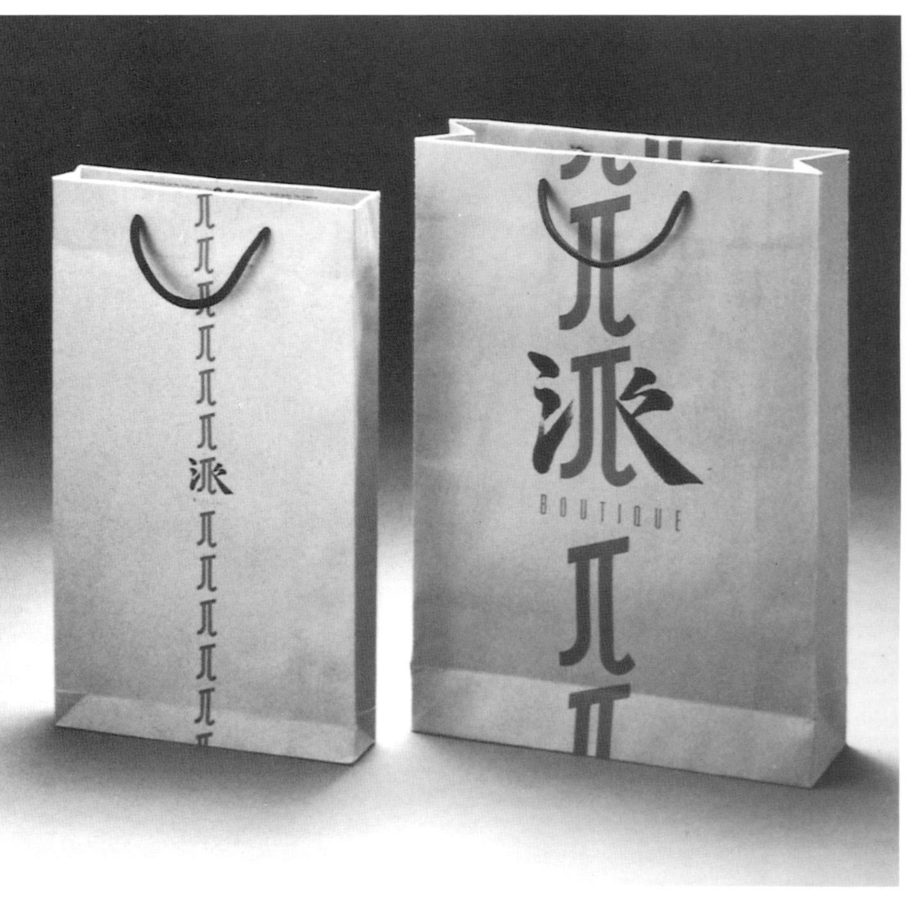

162

163

164

163
アパレル・ブティック
アトリエ伊藤佐智子
東京
1982

D／成瀬始子

164
個展のためのショッピングバッグ
東京デザイナーズスペース
東京
1987

D／川崎修司, 高橋新三
P／能津喜代房
DF／ヘッズ

165
婦人服店
阪神商事(アンドア)
神戸市
1988

D／草刈 順

165

163
Apparel Boutique
Atelier Ito Sachico
Tokyo, Japan
1982

D／Motoko Naruse

164
Shopping Bag Works for the One-man Exhibit
Tokyo Designer's Space
Tokyo, Japan
1987

D／Shuji Kawasaki, Shinzo Takahashi
P／Kiyofusa Nozu
DF／Heads Inc.

165
Ladies' Cothing Shop
Hanshinshoji Co., Ltd. (Andor)
Kobe City, Japan
1988

D／Jun Kusakari

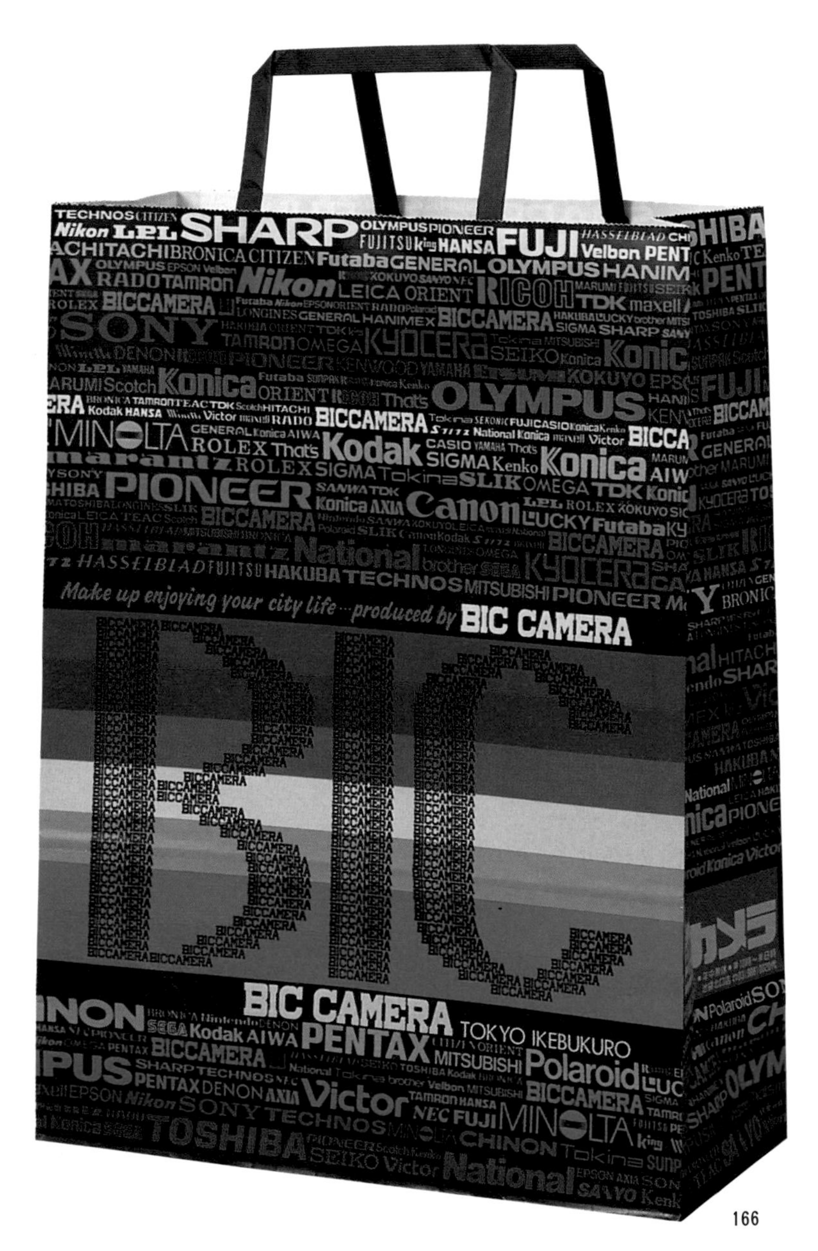

166

167

166
カメラ・家電製品ストアー
ビックカメラ
東京
1987

D／村越幸子
DF／ジャイア

167
磁気テープ製造会社
紙，ビニール
TDK
東京
1986

AD／和泉賢二
D／松本 隆
DF／TDKデザインコア
新しく展開中のBI戦略の一環として，より広くマークを訴求し，ブランドイメージを向上させることを目的としてデザインされた。

166
Camera & Electric Appliances Store
Bic Camera Co., Ltd.
Tokyo, Japan
1987

D／Sachiko Murakoshi
DF／Jia Co., Ltd.

167
Magnetic Tape Manufacturer
Paper, Plastic
TDK Corporation
Tokyo, Japan
1986

AD／Kenji Izumi
D／Takashi Matsumoto
DF／TDK Design Core Co., Ltd.
The Bags designed to improve the brand image by searching the potential power of the "TDK!" mark.

168
"USA-Time"
ビニール
ラインブリュッケ
スイス, バーゼル
1972

D/アーミン・フォクト
DF/アーミン・フォークト・パートナー

169
"バルツァース"
ビニール
バルツァース・アクティエンゲゼルシャフト
リヒテンシュタイン, フュルシュテントゥム
1985

AD/ゴットショーク+アッシュ・インターナショナル
D/フリッツ・ゴットショーク
DF/ゴットショーク+アッシュ・インターナショナル

170
"ブルクハルト・ラジオ・テレビジョン"
ビニール
ブルクハルト・ラジオ・テレビジョン
スイス, チューリッヒ
1987

AD/ゴットショーク+アッシュ・インターナショナル
D/フリッツ・ゴットショーク, マッシモ・マッツィ
DF/ゴットショーク+アッシュ・インターナショナル

168
"USA-Time"
Plastic
Rheinbrücke AG
Basel, Switzerland
1972

D/Armin Vogt
DF/Armin Vogt Partner

169
"Balzers"
Plastic
Balzers Aktiengesellschaft
Fürstentum, Liechtenstein
1985

AD/Gottschalk+Ash Int'l
D/Fritz Gottschalk
DF/Gottschalk+Ash Int'l

170
"Burkhardt AG Radio-Television"
Plastic
Burkhardt AG Radio-Television
Zürich, Switzerland
1987

AD/Gottschalk+Ash Int'l
D/Fritz Gottschalk, Massimo Mazzi
DF/Gottschalk+Ash Int'l

168

169

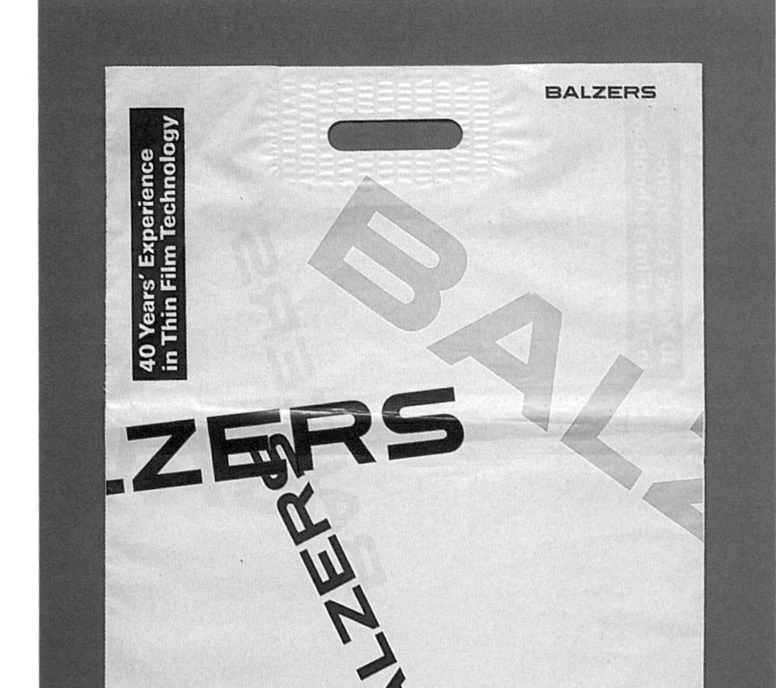

170

171

172

171
ブティック
ロック座
東京
1986

D／中山祐子
DF／ロック座

172
環境キャンペーン
ASNU(フィレンツェ市廃棄物収集局)
イタリア，フィレンツェ
1986

D／レオナルド・バグリオーニ
DF／スタジオ・レオナルド・バグリオーニ
フィレンツェで行なわれた環境キャンペーン・シリーズのひとつ，ペーパー・コレクション・キャンペーンのプロモーション用バッグ。

173
デパートの食料品用ショッピングバッグ
紙，ポリエチレン
松坂屋
東京
1970

D／松坂屋(本社営業本部)

171
Boutique
Rock-za Co., Ltd.
Tokyo, Japan
1986

D／Yuko Nakayama
DF／Rock-za Co., Ltd.

172
Environmental Campaign
ASNU (Municipal Garbage Collection Agency)
Firenze, Italy
1986

D／Leonardo Baglioni
DF／Studio Leonardo Baglioni
This bag was a promotional element in the paper collection campaign, one of a series of environmental campaigns is the city of Firenze (Florence).

173
Department Store Shopping Bag for Foods
Paper, Plastic
Matsuzakaya Co., Ltd.
Tokyo, Japan
1970

D／Matsuzakaya Co., Ltd.

173

174
化粧品会社
ヤクルト化粧品
東京
1987

AD／榊原勝一
D／碓井まり
DF／マリーパック

175
外国書籍販売会社
丸善
東京
1952(包装紙),
1986(ショッピングバッグ)

D／平野新二，山崎達雄
I／平野新二
DF／丸善宣伝課

174
Cosmetics Company
Yakult Cosmetics Co., Ltd.
Tokyo, Japan
1987

AD／Katsuichi Sakakibara
D／Mari Usui
DF／Marry Pack Co., Ltd.

175
Foreign Book Dealer
Maruzen Co., Ltd.
Tokyo, Japan
1952 (Wrapping Paper),
1986 (Shopping Bag)

D／Shinji Hirano , Tatsuo Yamazaki
I／Shinji Hirano
DF／Maruzen Co., Ltd.

174

175

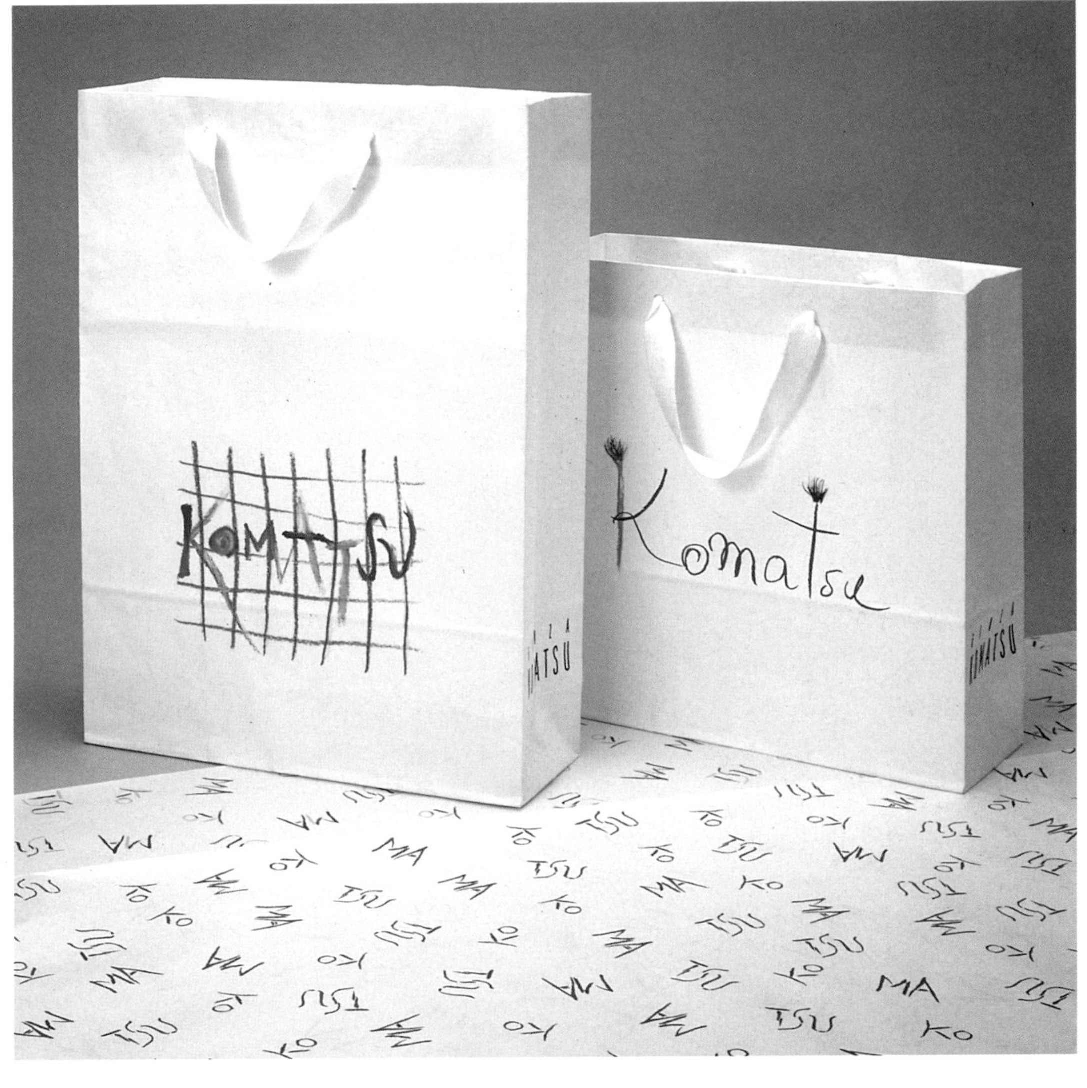

176

177

176
デパート
銀座小松
東京
1986

AD／長友啓典
D／長友啓典，黒田征太郎，野村高志
DF／K2

177
ファッションビルのクリスマス・クッキーバッグ
心斎橋 Vivre 21
大阪
1987

AD／永井ヒロ子
D／服部優里
DF／森久第二事務所
クリスマスに売り出したオリジナルクッキーのためのバッグ。パラフィン処理と5種類の手押しスタンプで，オリジナルの味を表現。

176
Department Store
Ginza Komatsu
Tokyo, Japan
1986

AD／Keisuke Nagatomo
D／Keisuke Nagatomo, Seitaro Kuroda, Takashi Nomura
DF／K-Two Co., Ltd.

177
Christmas Cookie Bag for Boutique Shopping House
Shinsaibashi Vivre 21
Osaka, Japan
1987

AD／Hiroko Nagai
D／Yuri Hattori
DF／Morihisa Second Co., Ltd.
A bag for an original brand of cookies promoted for the Christmas season. Creating a mood of unique original brand tastes by the introduction of the paraffin treatment method and five kinds of marks.

178・179
靴製造販売会社
フラテッリ・ロセッティ
アメリカ, ニューヨーク
1984

D／マッシモ・ヴィネリ
DF／ヴィネリ・アソシエイツ
イタリア製靴の製造/販売業者, フラテッリ・ロセッティが開発したC. I. プログラムの一環としてつくられた箱とバッグ。箱のサイドには同社のロゴがずらりと対角線状に並んだストライプがあしらわれているが, これはバッグのデザインにも見ることができる。

180
化粧品会社'87夏のキャンペーン
ポリエチレン
コーセー化粧品
東京
1987

D／今村倶子
P／中島敏夫
DF／コーセー化粧品

178・179
Shoe Manufacturers and Retailers
Fratelli Rossetti
New York, USA
1984

D／Massimo Vignelli
DF／Vignelli Associates
The boxes and the bags created for Fratelli Rossetti, manufacturer and retailer of fine Italian shoes, are part of a larger corporate identity program. The boxes have diagonal stripes on the sides made by repeating the Rossetti logo. The same repetitions occur on the paper bag.

180
Cosmetics Company '87 Summer Campaign
Plastic
Kose Cosmetics Co., Ltd.
Tokyo, Japan
1987

D／Tomoko Imamura
P／Toshio Nakajima
DF／Kose Cosmetics Co., Ltd.

178

179

180

181

182

181
印刷会社
光村原色版印刷所
東京
1987

AD／柳崎正雄
D／山田政彦
P／安西秀樹
DF／光村原色版印刷所
情報化社会に「清新で，軽快に即応する企業」というイメージを再構築できるデザインが制作意図。白地に曲線を描く青帯は，白紙が輪転機の中を運ばれながら印刷される様子を象徴し，表現意図には，清新さと軽快さ，そして至高への願望も。

182
化粧品会社'85冬のキャンペーン
コーセー化粧品
東京
1985

D／今村倶子
P／中島敏夫
DF／コーセー化粧品
テーマ商品のトリートメントを訴求するため，しっとり感とデリケートさがねらい。グレーと紫で色分割し，ロゴをソフトパープルでデザインポイントにしている。

181
Printing Company
Mitsumura Printing Co., Ltd.
Tokyo, Japan
1987

AD／Masao Yanasaki
D／Masahiko Yamada
P／Hideki Anzai
DF／Mitsumura Printing Co., Ltd.
Aiming to make a design which reestablishes the company in the information society, with the image of a "young, fresh business, ready to promptly meet current trends". A blue band curving against the white background stands for flow of white paper being passed through a printing press. Intended to present a fresh and light feeling, and even perhaps a desire for supremacy.

182
Cosmetics Company '85 Winter Campaign
Kose Cosmetics Co., Ltd.
Tokyo, Japan
1985

D／Tomoko Imamura
P／Toshio Nakajima
DF／Kose Cosmetics Co., Ltd.
A touch of moisture and delicacy is sought to emphasize the excellence of the theme treatment product. A design using the logo in light purple against the background in two-tone color of gray and purple is introduced.

183
ショッピングセンター
池袋ショッピングパーク
東京
1982

D／草刈 順

184
製パン会社
松月堂製パン
山口県宇部市
1982

D／野口正治
DF／スタジオ・コムテック

183
Shopping Center
Ikebukuro Shopping Park Co., Ltd.
Tokyo, Japan
1982

D／Jun Kusakari

184
Bakery
Shogetsudo Bakery Co., Ltd.
Ube City, Japan
1982

D／Masaharu Noguchi
DF／Studio Comtech

183

184

185

185
テイクアウトのそうざい屋
ポリエチレン
若菜
東京
1987

AD／田中一光
D／木下勝弘
DF／デザイン倶楽部

185
Side Dish Take-out Service Shop
Plastic
Wakana Co., Ltd.
Tokyo, Japan
1987

AD／Ikko Tanaka
D／Katsuhiro Kinoshita
DF／Design Club Co., Ltd.

186

187

186
ファッション・ブランド"プランテーション"
ビニール
イッセイミヤケインターナショナル
東京
1981

AD／成瀬始子
D／成瀬始子, 河上妙子
"光と大気と大地"をコンセプトとするシンボルマークをストレートにあしらい, 大地に浮かぶ風船のイメージをダイナミックな形状で表現している。背負うことができる。

187
ファッションビル"バサラハウス・チハラ"
ビニール
チハラ
大分市
1985

D／成瀬始子
大袋は, たっぷりとした肩紐で, ダイナミックな持ち姿を想定してデザインしている。

186
Fashion Brand "Plantation"
Plastic
Issey Miyake International Inc.
Tokyo, Japan
1981

AD／Motoko Naruse
D／Motoko Naruse, Taeko Kawakami
A symbol based on the concepts of "Light, Air and Earth", expressed in the dynamic shape of a ballon image, rising up from the Earth. This bag can also be used as a ruck sake.

187
Boutique Shopping House
"Basara House Chihara"
Plastic
Chihara Co., Ltd.
Oita City, Japan
1985

D／Motoko Naruse
This large bag has a long shoulder strap and has been designed to look dynamic when being carried.

188

188
洋酒会社のギフト用バッグ
サントリー
大阪
1986

D／前田英樹
DF／サントリー・デザイン部

189
洋菓子会社のバレンタイン・デー
用バッグ
モロゾフ
神戸
1987

D／高橋 篤

188
Liquor Company Gift Bag
Suntory Limited
Osaka, Japan
1986

D／Tsuneki Maeda
DF／Suntory Limited

189
Valentine's Day Shopping Bag
for Confectionery
Morozoff Limited
Kobe City, Japan
1987

D／Atsushi Takahashi

189

190

191

190
スポーツウェア・ショップ
ビニール
インタークルー
名古屋市
1987

D／岡本 昭
DF／インタークルー

191
書店
ビニール
ブッホハンドルンク・シェルツ
スイス, ベルン
1979

D／クルト・ヴィルト

192
デパートのオリジナル・ブランド"Be Living"
小田急百貨店
東京
1982

D／松永 真
このデパートのインテリア家具から食料品までを，一貫した生活理念のもとに統合するためにつくられたブランド"Be Living"のショッピングバッグ。

190
Sporting Wear Shop
Plastic
Intercrew
Nagoya City, Japan
1987

D／Akira Okamoto
DF／Intercrew

191
Bookstore
Plastic
Buchhandlung Scherz
Bern, Switzerland
1979

D／Kurt Wirth

192
Department Store Original Brand "Be Living"
Odakyu Department Stores Co., Ltd.
Tokyo, Japan
1982

D／Shin Matsunaga
The shopping bag for "Be Living" brands was designed with the aim of bringing together ideas of different lifestyles, ranging from furniture to food stuffs.

192

193

193
美術館
紙, ビニール
ロイヤル・オンタリオ美術館
カナダ, トロント
1970

D／バートン・クレイマー
DF／バートン・クレイマー・アソシエイツ

193
Museum
Paper, Plastic
Royal Ontario Museum
Toronto, Canada
1970

D／Burton Kramer
DF／Burton Kramer Associates Ltd.

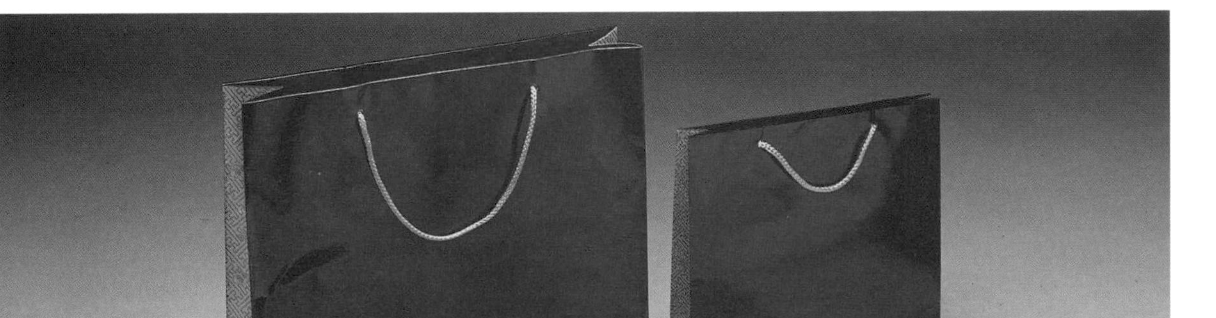

194

194
健康食品ストアー
ヘルス・フード・ストアー
ホンコン
1982

AD／カン・タイ-クン
D／カン・タイ-クン, リー・サイ・チュー
DF／SS デザイン・アンド・プロダクション
ヘルス・フード・ストアー社は漢方薬, および海産物を扱うホンコンの卸し・販売業者。中国特有の黄と赤の模様の上に「補品店」のロゴがバッグの両面に入っている。

194
Health Food Store
Health Food Store Ltd.
Hong Kong
1982

AD／Kan Tai-keung
D／Kan Tai-keung, Li Sai Chiu
DF／SS Design & Production
Health Food Stores Limited is a wholesaler and retailer of Chinese drugs nad marine products in Hong Kong. The golden Longevity Chinese pattern, yellow and red color as the background, the logotype of Chinese store name is printed on both-side of the bag, which looks most oriental.

195

196

195
宝石店
ヴァネッサ・ジュエリー
ホンコン
1982

AD／カン・タイークン
D／カン・タイークン, リ・サイ・チュー
DF／SS デザイン・アンド・プロダクション
この宝石店のショッピングバッグの金と黒の色彩は，宝石のイメージを強調したもの。"ヴァネッサ"というロゴが高級感をそえている。

195
Jewelry Store
Vanessa Jewelry
Hong Kong
1982

AD／Kan Tai-keung
D／Kan Tai-keung, Li Sai Chiu
DF／SS Design & Production
Gold and black colors of this jewelry store shopping bag emphasize the image of jewelry. The name "Vanessa" is set in type that adds a sense of elegance.

196
現代美術センター内書店
シンシナティ現代美術センター
アメリカ, シンシナティ
1983

AD／チャック・バイアン
D／チャック・バイアン, マイケル・オーバトーン
DF／チャック・バイアン・デザイン
小さなバッグのグラフィックが，故意にわかりづらくデザインされている。

196
Contemporary Arts Center Bookstore
Cincinnati Contemporary Arts Center
Cincinnati, USA
1983

AD／Chuck Byrne
D／Chuck Byrne, Michael Overton
DF／Chuck Byrne Design
The graphic on the small bag was intended to be difficult to read.

197
ファッション・ブランド"タクティクス・デザイン"
ザ・ギンザ
東京
1983

D／仲條正義

197
Fashion Brand "Tactics Design"
Shiseido Boutique The Ginza
Tokyo, Japan
1983

D／Masayoshi Nakajo

197

198
"Intel '87"
バッサーニ・ティチーノ
イタリア，ミラノ
1987

D／ハインツ・ウァイブル
DF／シグノ SRL

198
"Intel '87"
Bassani Ticino
Milano, Italy
1987

D／Heinz Waibl
DF／Signo SRL

199
ブティック
ワンダーハウス
東京
1983

AD／玉井 博
D／高橋明宏
DF／リムジン・インターナショナル

199
Boutique
Wonder House
Tokyo, Japan
1983

AD／Hiroshi Tamai
D／Akihiro Takahashi
DF／Limousine Internatinal Corp.

200
"キタイ"
キタイ
イタリア，コモ
1976

D／ウォルター・バルマー
DF／ウニデザイン

200
"Kitaj"
Kitaj
Como, Italy
1976

D／Walter Ballmer
DF／Unidesign

198

199

200

201
美術大学
ビニール
嵯峨美術短期大学
京都
1986

D／鯛天成雄

201
Art College
Plastic
Saga Junior College of Art
Kyoto, Japan
1986

D／Nario Taiten

201

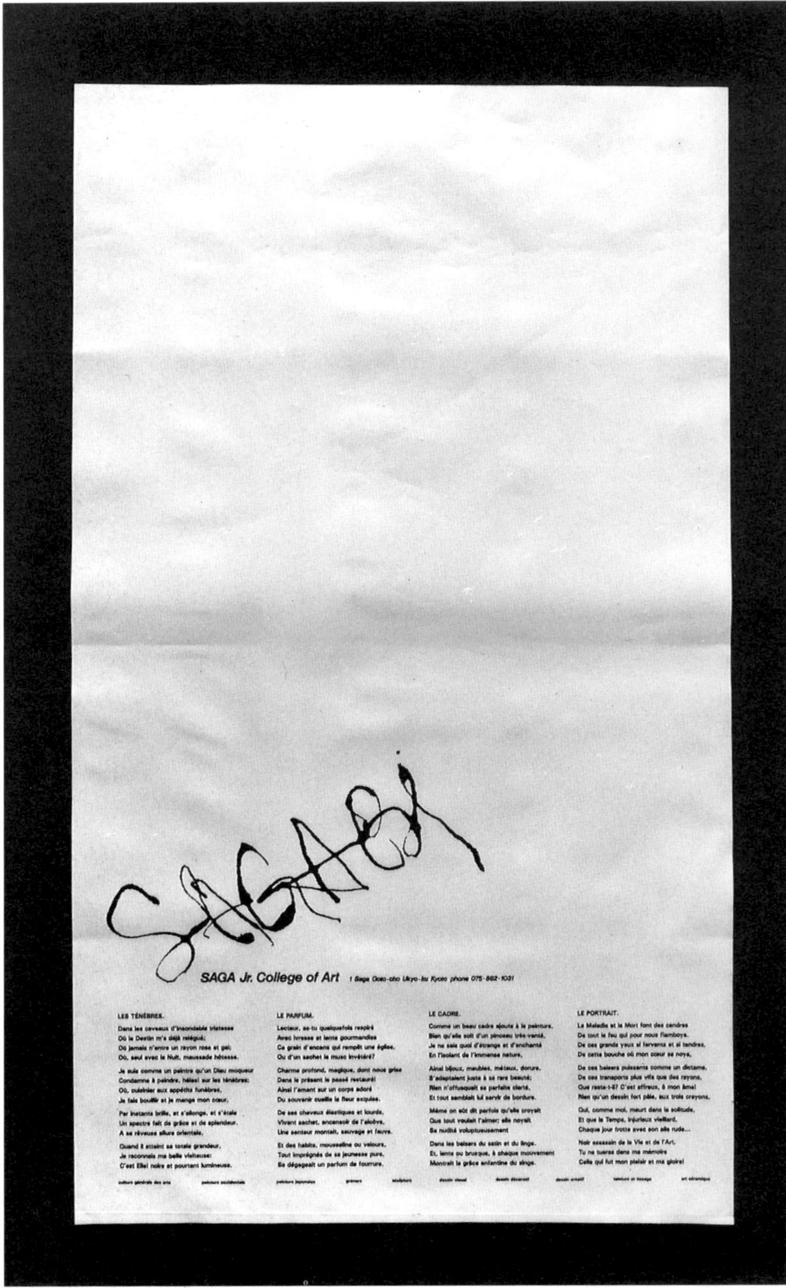

202

202
スウェーデン・インド祭1987
インド手織物・手工芸品輸出会社
インド, ニューデリー
1987

D／ベノイ・サーカー
DF／HHEC・デザイン室
スウェーデンで行なわれたインド祭でのみやげもの用につくられたクラフト紙の紙袋。

202
Indian Festival ; Sweden 1987
Handlooms & Handicrafts Export Corp. of India Ltd.
New Delhi, India
1987

D／Benoy Sarkar
DF／HHEC Design Cell
This brown paper bag was produced for The Festival of India in Sweden and used for souvenir items.

203
子供服ブティック
キムラタン(ミクサージュ)
神戸市
1986

AD/井口雅文
D/近藤真由美
I/藤倉良行
DF/イマジネ

204
靴店
カルロス・ファルチ・ジャパン
東京
1986

AD/カルロス・ファルチ
DF/ドラゴンウィック・デザインズ・ニューヨーク

203
Children's Clothing Shop
Kimuratan Co., Ltd. (Mix Age)
Kobe City, Japan
1986

AD/Masafumi Iguchi
D/Mayumi Kondo
I/Yoshiyuki Fujikura
DF/Imagine

204
Shoe Store
Carlos Falchi Japan Co., Ltd.
Tokyo, Japan
1986

D/Carlos Falchi
DF/Dragonwyck Designs Ltd., in N.Y.

203

204

205

205
デパート
高島屋
東京
1986

AD／安部修二
D／早川博唯
DF／TAD

205
Department Store
Takashimaya Co., Ltd.
Tokyo, Japan
1986

AD／Shuji Abe
D／Hirotada Hayakawa
DF／TAD Co., Ltd.

206
ブティック
ビニール
ノース・マリン・ドライブ
東京
1987

D／ノース・マリン・ドライブ

207
アパレル・ブティック
ビニール
ATTIC
神戸市
1987

D／ATTIC
DF／キモリ産業

208
レディース・ブティック
ティ・ファニー
名古屋市
1987

D／加藤多恵子

206
Boutique
Plastic
North Marine Drive Co., Ltd.
Tokyo, Japan
1987

D／North Marine Drive Co., Ltd.

207
Apparel Boutique
Plastic
ATTIC
Kobe City, Japan
1987

D／ATTIC
DF／Kimori Sangyo Co., Ltd.

208
Ladies Boutique
Ti Ffany
Nagoya City, Japan
1987

D／Taeko Kato

206

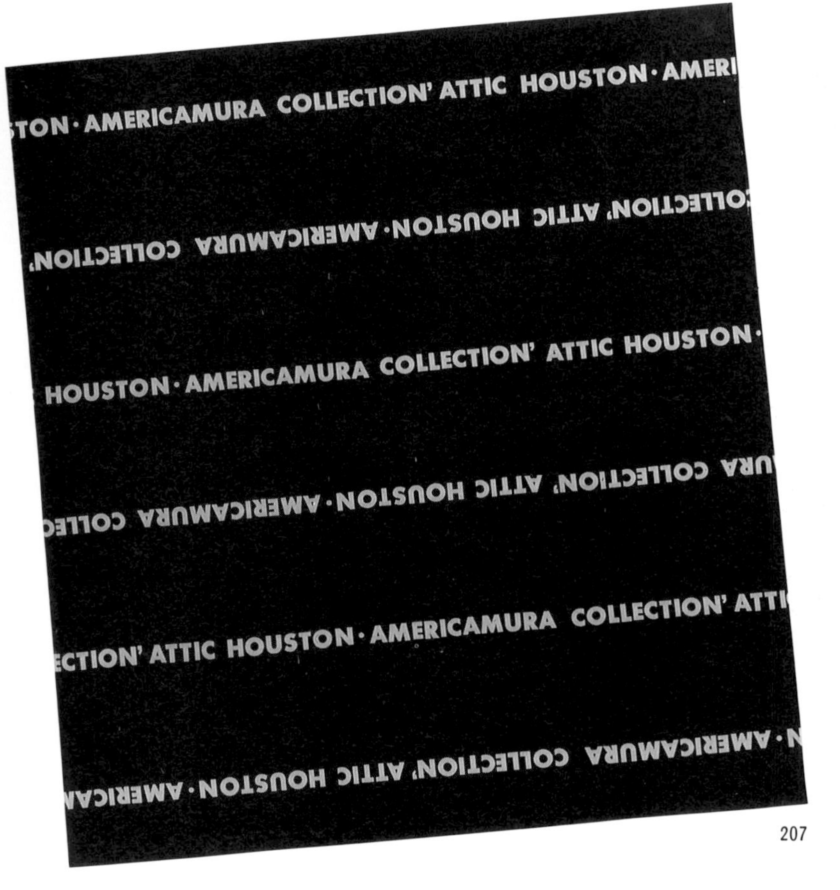

207

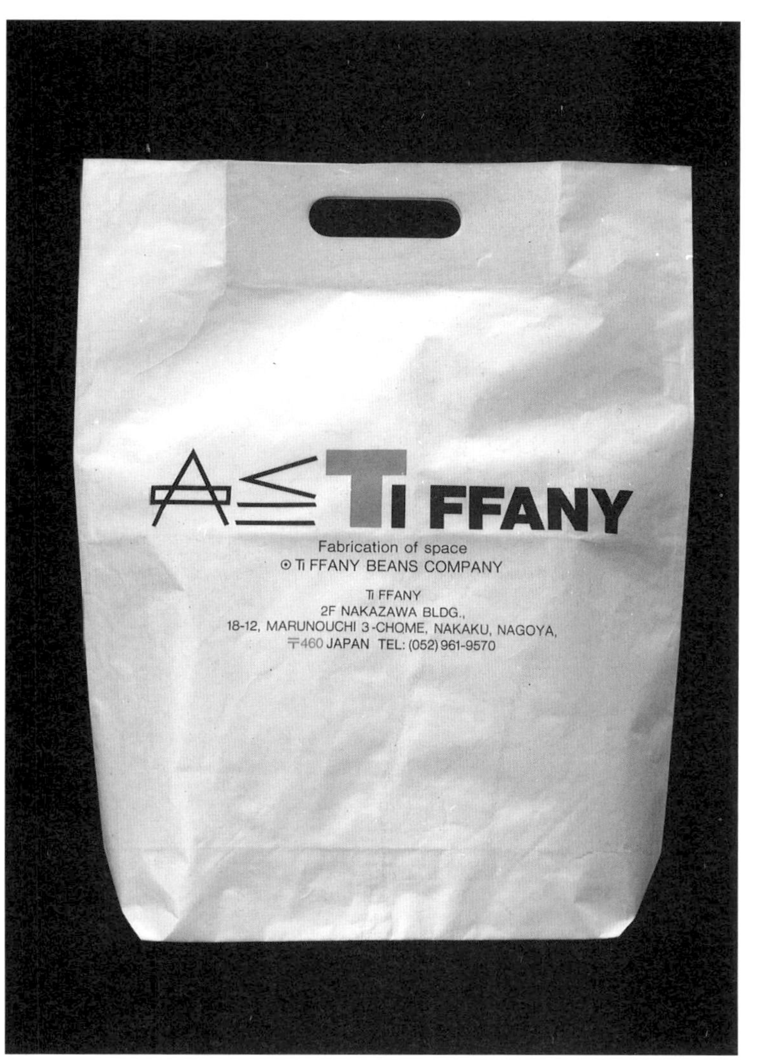

208

209

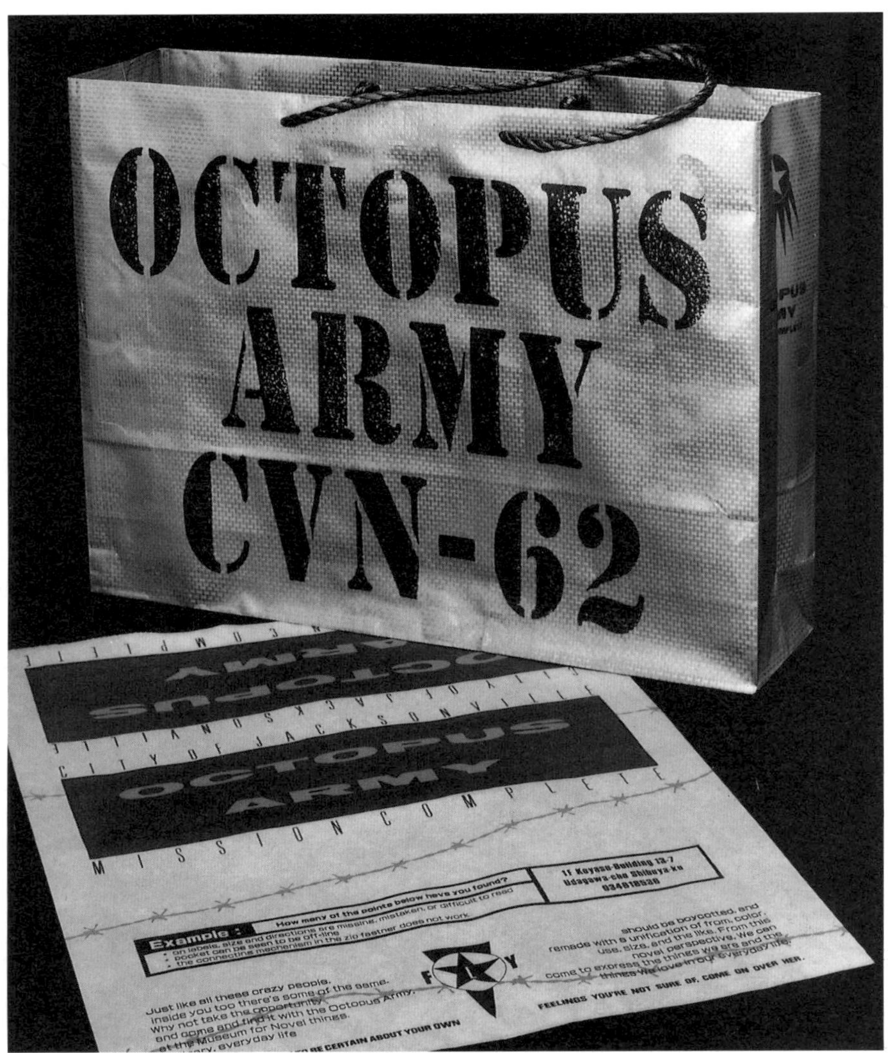

210

209
スポーツ用品会社
ビニール
YASSA
ユーゴスラヴィア，ヴァラスディン
1985

D／ボリス・リュービチック
DF／スタジオ・インターナショナル
YASSA のロゴは，ユーゴスラヴィアの国旗のデザインにもとづいている。このスポーツ用品会社の重要な販売ツールとなっているビニール・バッグ。

210
アパレル・ブティック
デライツ・クラブ（オクトパス・アーミー）
東京
1984

AD／速見敏昭
D／速見牧子
DF／夢幻

209
Sporting Goods Maker
Plastic
YASSA
Varaždin, Yugoslavia
1985

D／Boris Ljubičić
DF／Studio International
A／CIO, Zagreb
YASSA's logotype is derived from the official flag of Yugoslavia. This plastic bag is one of the sporting goods maker's most important promotinal articles.

210
Apparel Boutique
D'Lites Club (Octpus Army)
Tokyo, Japan
1984

AD／Toshiaki Hayami
D／Makiko Hayami
DF／Mugen Corporation

211

211
テナントビルの夏のキャンペーン
岡田屋モアーズ
横浜・川崎
1984

AD／杉本英介
D／野本卓司
DF／ブレックファースト

212
ブティック
グランド・キャニオン(ムパタ)
東京
1986

D／和井内京子
DF／テレビ朝日ミュージック・杉田事務所

211
Boutique Shopping House
Okadaya-More's
Yokohama, Japan
1984

AD／Eisuke Sugimoto
D／Takuji Nomoto
DF／Breakfast Co., Ltd.

212
Boutique
Grand Canyon Co., Ltd.
(Mpata Shop)
Tokyo, Japan
1986

D／Kyoko Wainai
DF／Sugita Office, T.V.Asahi Music Co., Ltd.

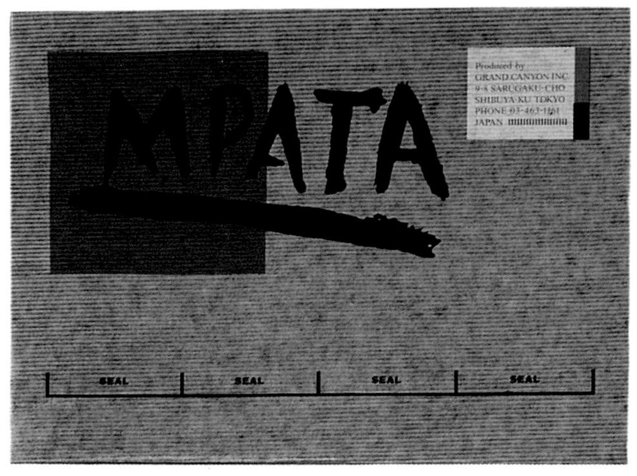

212

213—218
個展のためのショッピングバッグ
東京デザイナーズスペース
東京
1987

D／川崎修司, 髙橋新三
P／能津喜代房
DF／ヘッズ
東京デザイナーズスペースでの個展のために，取引のある10社を選んで自由に制作したもの。

213—218
Shopping Bag Works for the One-man Exhibit
Tokyo Designer's Space
Tokyo, Japan
1987

D／Shuji Kawasaki, Shinzo Takahashi
P／Kiyofusa Nozu
DF／Heads Inc.
Optionally designed works on selected items for ten clients for one-man exhibiting at the Tokyo Designer's Space.

213

214

215

216

217

218

219
アパレル・ブティック
デライツ・クラブ(リトル・スリーク・ビーン)
東京
1984

D／速見牧子
DF／夢幻

220
アパレル・ブティック
デライツ・クラブ(イエロー・モンスター)
東京
1984

D／速見牧子
DF／夢幻

219
Apparel Boutique
D'Lites Club (Little Sleek Bean)
Tokyo, Japan
1984

D／Makiko Hayami
DF／Mugen Corporation

220
Apparel Boutique
D'Lites Club (Yellow Monster)
Tokyo, Japan
1984

D／Makiko Hayami
DF／Mugen Corporation

219

220

221

221
市販用"レディー"
シモジマ商事
東京
1987

I／灘本唯人
DF／シモジマ商事

222
アパレル・ブティック
ビニール
NYLON
大阪

AD／NYLON
D／辻本仁史

223
バラエティ・ショップ
ビニール
ポップ・ショップ・エンタープライズ
(Pop Shop Tokyo)
東京
1988

I／キース・ヘリング

221
Shopping Bag "Lady"
Shimojima Shoji Co., Ltd.
Tokyo, Japan
1987

I／Tadahito Nadamoto
DF／Shimojima Shoji Co., Ltd.

222
Apparel Boutique
Plastic
NYLON
Osaka, Japan

AD／NYLON
D／Hitoshi Tsujimoto

223
Variety Shop
Plastic
Pop Shop Enterprise Co., Ltd.
(Pop Shop Tokyo)
Tokyo, Japan
1988

I／Keith Haring

NYLON
NYLON
NYLON
NYLON

222

POP
SHOP
TOKYO
TM
©K.HARING⊕

223

POP
SHOP
TOKYO
TM
©K.Haring 87⊕
3-11-12 MINAMIAOYAMA MINATO-KU TOKYO〒107
TEL.-03-408-0207

224

225

224
新聞整理用の紙袋
毎日新聞社
東京
1987

AD／石崎洋之
D／田原 香
I／佐川みどり
DF／ヴァーナル・アンド・カンパニー
A／スーパーバッグ S.P.C. 室

225
化粧品会社'81夏のキャンペーン
資生堂
東京
1981

AD／中山禮吉
D／伊藤 満
P／小暮 徹

224
Paper Bag for Old Newspapers
Mainichi Shimbunsha Inc.
Tokyo, Japan
1987

AD／Hiroyuki Ishizaki
D／Kaori Tahara
I／Midori Sagawa
DF／Vernal And Company Ltd.
A／S.P.C.Room, Super Bag Co., Ltd.

225
Cosmetics Company '81 Summer Campaign
Shiseido Co., Ltd.
Tokyo, Japan
1981

AD／Reikichi Nakayama
D／Mitsuru Ito
P／Tohru Kogure

226
国際ボートショー
日本舟艇工業会
東京
1987

AD／武下 朗
D／北村明倫子
DF／武下 CD 事務所
A／電通

227
化粧品会社の夏のキャンペーン
資生堂
東京
1980

AD／田島一夫
D／安原和夫
DF／資生堂宣伝部

226

226
International Boat Show
Japan Boating Industry Assoc.
Tokyo, Japan
1987

AD／Akira Takeshita
D／Akiko Kitamura
DF／Takeshita Creative Director Co., Ltd.
A／Dentsu, Inc.

227
Cosmetics Company Summer Campaign
Shiseido Co., Ltd.
Tokyo, Japan
1980

AD／Kazuo Tajima
D／Kazuo Yasuhara
DF／Shiseido Co., Ltd.

227

228

229

228
デパートのクリスマスセール
松坂屋(上野店)
東京
1985

D/久保田恭博
DF/松坂屋(上野店)

229
デパート
西武百貨店
東京
1986

I/大西洋介
DF/田中一光デザイン室

228
Department Store Christmas Sale
Matsuzakaya Co., Ltd. (Ueno Store)
Tokyo, Japan
1985

D/Yasuhiro Kubota
DF/Matsuzakaya Co., Ltd. (Ueno Store)

229
Department Store
The Seibu Department Stores Co., Ltd.
Tokyo, Japan
1986

I/Yosuke Onishi
DF/Ikko Tanaka Design Studio

230
ファッションビルのクリスマスセール
大宮ステーションビル
埼玉県大宮市
1986—1987

D/八木健夫
I/柳田 亘
DF/オフィース・ピーアンドシー

231
デパート
高島屋
東京
1986

AD/二村恒典
D/沼田博美
I/ビヤン・ヴィンブランド
DF/宣研

232
デパート
高島屋
東京
1967

AD/高崎敏武
D/姫井 稔
P/後藤邦夫
DF/宣研

230
Shopping Center Christmas Sale
Omiya Station Building
Omiya City, Japan
1986—1987

D/Takeo Yagi
I/Wataru Yanagida
DF/Office P&C Inc.

231
Department Store
Takashimaya Co., Ltd.
Tokyo, Japan
1986

AD/Tsunenori Nimura
D/Hiromi Numata
I/Bjorn Wiinblad
DF/Senken Co., Ltd.

232
Department Store
Takashimaya Co., Ltd.
Tokyo, Japan
1967

AD/Toshitake Takasaki
D/Minoru Himei
P/Kunio Goto
DF/Senken Co., Ltd.

230

231

232

233

234

233
革細工店
わち・ふぃーるど
東京
1984

AD／池田晶子
D／池田晶子, 幸 英子, 土屋ゆみ子
DF／わち・ふぃーるど

234
ブティック
タック（コンガ）
東京
1987

AD／長田 章
D／長田 章, 岩見伸一
I／林 幸蔵
DF／オサダ
A／東京アートパッケージ

233
Leathercraft Shop
Wachi Field Co., Ltd.
Tokyo, Japan
1984

AD／Akiko Ikeda
D／Akiko Ikeda, Fusako Yuki, Yumiko Tsuchiya
DF／Wachi Field Co., Ltd.

234
Boutique
Tac Corporation (Konga)
Tokyo, Japan
1987

AD／Akira Osada
D／Akira Osada, Shinichi Iwami
I／Kozo Hayashi
DF／Osada & Co., Ltd.
A／Tokyo Art Package Co., Ltd.

235

235
ファッション・ブティック
ブティック・バザール
ホンコン
1982

AD／アラン・チャン
D／アラン・チャン, アルヴィン・チャン
DF／アラン・チャン・デザイン
"ブティック・バザール"はヨーロッパのファッション・ブランドを販売している香港の高級ブティック。

235
Fashion Boutique
Boutique Bazaar
Hong Kong
1982

AD／Alan Chan
D／Alan Chan, Alvin Chan
DF／Alan Chan Design Co.
Boutique Bazaar is one of Hong Kong's leading high fashion boutiques, selling designer apparel from Europe.

236

236
子供用品店
イマジネイリアム
アメリカ，カリフォルニア州ウォルナット・クリーク
1987

AD／マイケル・メイブリー，シドニー・キャメロン
D／マイケル・メイブリー，マーギー・チュー
I／N/A
DF／マイケル・メイブリー・デザイン
カリフォルニアにある一連の子供用品のショップのためのストアー・グラフィック・システムの一環としてつくられたショッピングバッグ。ショップ内には子供たちが自由に遊べるようにおもちゃやゲーム，本などが置かれている。バッグには動物たちや童話のキャラクターなどのイメージが浮き彫りにされている。

237
洋紙店の紙袋
井上洋紙店
大阪
1988

AD／奥村昭夫
D／大高郁子（a，d，f，h）
竹村真澄（b）
田中幹人（c）
野上周一（e）
切間晴美（g）
I／王 永華（h）
DF／パッケージング・クリエイト

236
Children's Store
Imaginarium
Walnut Creek, CA, USA
1987

AD／Michael Mabry, Sydney Cameron
D／Michael Mabry, Margie Chu
I／N/A
DF／Michael Mabry Design
The shopping bag is part of the store graphics system for a series of children's store located in California. The stores are participatory in nature, where children can play with all the toys, games, books, etc. in the store. The bag combines engraved images of animals, fairy tale characters, and elements in nature.

237
Paper Bag for Paper Dealer
Inoue Yoshiten Co., Ltd.
Osaka, Japan
1988

AD／Akio Okumura
D／Ikuko Ohtaka (a, d, f, h)
Masumi Takemura (b)
Mikito Tanaka (c)
Shuichi Nogami (e)
Kirima Harumi (g)
I／Eika Oh (h)
DF／Packaging Create Inc.

INOUE
PA·PERS
Pretty as a Picture
POEM
FIVE BIRDS
Where is one more bird ?
AMAZING MONSTER
STRANGE ADVENTURE ON OTHER WORLD

a b c d e h f g

238
“サッサン”
サッサン・コーポレーション・リミテッド
オーストラリア, メルボルン
1986

D／ケン・ケイトー
DF／ケイトー・ デザイン・インコーポレイテッド

238
“Sussan”
Sussan Corporation (Australia) Ltd.
Melbourne, Australia
1986

D／Ken Cato
DF／Cato Design Inc.

239
展示会用のサービスバッグ
スーパーバッグ
東京
1987

AD／スーパーバッグ S.P.C. 室
D／石崎洋之
I／エリン・フィッツパトリック
DF／ヴァーナル・アンド・カンパニー

239
Service Bag for an Exhibition
Super Bag Co., Ltd.
Tokyo, Japan
1987

AD／S.P.C.Room, Super Bag Co., Ltd.
D／Hiroyuki Ishizaki
I／Erin Fitzpatrick
DF／Vernal And Company Ltd.

240
ブティック
ビニール
バルーン
東京
1985

D／安達澄江

241
ぬいぐるみ店
アクロス(ami)
東京
1982

AD／大月真由美
D／望月由美子
I／堀尾英子
DF／アクロス

242
ファッションビル
立川 WILL
東京
1987

AD／中山 寿
D／橋本良子
I／舟橋全二
DF／オリコミ
女性客の多いファッションビルの5周年記念のためのショッピングバッグで，若い女性の顔をシンプルな形で表わし，明るいイメージを意図。

240
Boutique
Plastic
Balloon Co., Ltd.
Tokyo, Japan
1985

D／Sumie Adachi

241
Stuffed Toy Shop
Across Co., Ltd. (ami)
Tokyo, Japan
1982

AD／Mayumi Otsuki
D／Yumiko Mochizuki
I／Eiko Horio
DF／Across Co., Ltd.

242
Shopping Center
Tachikawa WILL
Tokyo, Japan
1987

AD／Hisashi Nakayama
D／Yoshiko Hashimoto
I／Zenji Funabashi
DF／Orikomi Advertising Ltd.
The shopping bag commemorating the fifth anniversary of a shopping center, whose main customers are women. A simple design using the face of a young woman projects a bright, stylish image.

241

242

243

244

243
アパレル・ブティック
カレン・チャールズ
アメリカ, コネチカット州エンフィールド
1987

AD／ロバート・ガーシン
D／スコット・ボレストリッジ
I／ジョアン・ホール
DF／ロバート・ガーシン・アソシエイツ
“カレン・チャールズ”はアダルトな女性をターゲットにしたチェーン・ブティック。これは店内改装を含むトータル・グラフィック・プログラムの一環としてつくられたバッグである。

244
デパート
ハロッズ・ナイトブリッジ・リミテッド
イギリス, ロンドン

D／ミナーレ・タタースフィールド・プラス・パートナーズ

243
Chain of Clothing Stores
Caren Charles
Enfield, CT, USA
1987

AD／Robert Gersin
D／Scott Bolestridge
I／Joan Hall
DF／Robert P. Gersin Associates, Inc.
This shopping bag was created for Caren Charles, a chain of clothing stores featuring apparel for the mature woman. The bag is part of a total graphic and store redesign program.

244
Department Store
Harrods Knightbridge Ltd.
London, England

D／Minale Tattersfield+ Partners

245
ブティック
リチャード(ビザール)
神戸市
1982

D/横尾忠則

245
Boutique
Richard Co., Ltd. (Bizarre)
Kobe City, Japan
1982

D/Tadanori Yokoo

245

246

246
デパート
ブルーミングデイルズ
アメリカ, ニューヨーク
1987

AD/ジョン・ジェイ
D/ティム・ガーヴィン
DF/ティム・ガーヴィン・デザイン

246
Department Store
Bloomingdale's
New York, USA
1987

AD/John Jay
D/Tim Girvin
DF/Tim Grirvin Design, Inc.

247

248

247
ブック・センター
アーバン・ブック・センター
アメリカ・ニューヨーク
1984

D／R・O・ブレックマン
DF／R・O ブレックマン・インコーポレイテッド
最初に描かれたイラストは，本を読んでいる人が木によりかかっている構図だったが，後にクライアントの意向で木がビルに変更された。アーバン・ブック・センターは建築に関する本を売っていることが，変更の理由であった。

248
雑貨店
ワンダー貿易（宇宙百貨）
大阪
1987

AD／和田允宏
I／石丸千里
DF／Art Wad's

249
市販用"エリート"
松城紙袋工業
大阪
1987

D／宇田 優
I／奥田一明
DF／ユウ・デザイン事務所
バラエティーショップや薬品店，化粧品店などに置く有料バッグ。数点のシリーズで制作し，シーズンやトレンドでデザインを変える。

250
市販用"Key Staff 21"
大和シュガル
東京
1984

AD／能勢行正
D／加賀屋雅子
DF／大和シュガル

247
Book Center
The Urban Book Center
New York, USA
1984

D／R. O. Blechman
DF／R. O. Blechman Inc.
The illustration, parodies the classic situation in bookplates where a reader is pictured seated on the ground leaning against a tree. The tree was changed into a skyscraper because the client, The Urban Book Center, sells books on architecture.

248
Grocery Store
Wonder Trading Co., Ltd.
(Uchu-Hyakka)
Osaka, Japan
1987

AD／Mitsuhiro Wada
I／Chisato Ishimaru
DF／Art Wad's

249

250

249
Shopping Bag "Elete"
Matsushiro Paper Bag Industry Co., Ltd.
Osaka, Japan
1987

D/Masaru Uda
I/Kazuaki Okuda
DF/U Design Office
A bag, available at drugstores, cosmetic stores, and variety shops. Several series have been produced, and can be changed according to the season or consumer trends.

250
Shopping Bag "Key Staff 21"
Sugal Creative Production
Tokyo, Japan
1984

AD/Yukimasa Nose
D/Masako Kagaya
DF/Sugal Creative Production

251

251

252

251
ワイン袋
サントリー
東京
1986

D／中崎宣弘
DF／サントリー・パッケージング部
ワイン用手さげ袋。白ワインと赤ワインのボトルをモチーフにそれぞれ1本ずつ両面に印刷されている。イラストは実物とほぼ同サイズ。楽しい気分をストレートに演出することを目的としてデザインされた。

252
製パン会社
アローム
大阪
1980

D／嶋 高宏
DF／嶋デザイン事務所
“おしゃれなパリ”のイメージを女性に親しまれるようにやさしいタッチの猫とポエムで表現。色，デザインとも，持ち歩くときのファッション性を考慮。

251
Wine Bottle Bag
Suntory Limited
Tokyo, Japan
1986

D／Nobuhiro Nakazaki
DF／Packaging Dept., Suntory Limited.
A bag for wine bottle. White wine and red wine bottles are used as the motif. Each panel of the bag bears each bottle in the actual size. The design aims at expressing a delightful feeling in a straightforward manner.

252
Bakery
Arome Co., Ltd.
Osaka, Japan
1980

D／Takahiro Shima
DF／Shima Design Office Inc.
Expressing the image of “fashionable Paris” with a cat with a gentle touch and a poem to capture a woman's mind. The color and design are selected with a view to fashionability on the street.

253
外食会社
グリーンハウス
東京
1985

AD／米田喜久男
D／新川みさ子
I／米田喜久男
DF／ドット
A／凸版印刷

254
フルーツ＆ファッション・ブティック
新宿高野
東京
1972

D／石岡瑛子

253
Food Service Company
Green House Co., Ltd.
Tokyo, Japan
1985

AD／Kikuo Yoneda
D／Misako Shinkawa
I／Kikuo Yoneda
DF／Dot Co., Ltd.
A／Toppan Printing Co., Ltd.

254
Fruit & Fashion Boutique
Shinjuku Takano Co., Ltd.
Tokyo, Japan
1972

D／Eiko Ishioka

253

254

255
ブティック
ビニール
ピーナッツ・ボーイ
東京
1987

D／房野 満

256
フォトスタジオ
楽園
東京
1983

D／中村政久

255
Boutique
Plastic
Peanut Boy
Tokyo, Japan
1987

D／Mitsuru Fusano

256
Photo Studio
Rakuen
Tokyo, Japan
1983

D／Masahisa Nakamura

255

256

257
アパレル・ブティック
Can
東京
1985

AD／藤井 浩
D／藤井美津江
I／佐藤文恵
DF／Can

257
Apparel Boutique
Can Co., Ltd.
Tokyo, Japan
1985

AD／Hiroshi Fujii
D／Mitsue Fujii
I／Fumie Sato
DF／Can Co., Ltd.

257

258

259

258
子供服店
ビニール
Kid's Monster
東京
1986

D／木島 努
DF／リフレックス

259
洋菓子会社のハロウィーン用バッグ
ポリエチレン
モロゾフ
神戸
1987

D／高橋 篤

260
子供服店
ディボア
東京
1986

D／酒本雅夫

258
Children's Clothing Shop
Plastic
Kid's Monster Inc.
Tokyo, Japan
1986

D／Tsutomu Kijima
DF／Reflex Inc.

259
Halloween Shopping Bag for Confectionery
Plastic
Morozoff Limited
Kobe City, Japan
1987

D／Atsushi Takahashi

260
Children's Clothing Shop
Diboa
Tokyo, Japan
1986

D／Masao Sakamoto

260

261・263
デパートの英国フェア・キャンペーン
三越
東京
1987

AD/安藤 亨
D/木本映右
I/ボルト・ナッツ・スタジオ
DF/三越デザイン開発室
A/電通
英国をシンボライズしたキャラクター"ベン・スコット"が，バッグの周囲をぐるりと回るデザイン。ポスター，新聞，その他のコミュニケーション・メディアと連動して，文字どおり"動く広告"としての相乗効果が求められた。

262
洋菓子店
ローゼン製菓
大阪
1982

AD/成瀬政敏
I/湯村輝彦
DF/パノラマ・アドバタイジング

264
デパートのクリスマス・キャンペーン
三越
東京
1987

AD/安藤 亨
D/高橋道春
DF/三越デザイン開発室

265
商店街
上野公園通り商店会
東京
1982

AD/湯村輝彦
D/藤原哲男
I/湯村輝彦
DF/フラミンゴ・スタジオ，ステレオ・スタジオ

261・263
Department Store British Fair Campaign
Mitsukoshi Ltd.
Tokyo, Japan
1987

AD/Tohru Ando
D/Eisuke Kimoto
I/Bolt & Nuts Studio
DF/Design Room, Mitsukoshi Ltd.
A/Dentsu, Inc.
"Mr. Ben Scott" is a character symbolizing Britain, and becomes the design going around this bag. It is used the same time on posters, in newspapers and other communications media, and has an exponential effect through being "advertising that moves."

262
Confectionery
Rosen Co., Ltd.
Osaka, Japan
1982

AD/Masatoshi Naruse
I/Teruhiko Yumura
DF/Panorama Advertising Inc.

264
Department Store Christmas Campaign
Mitsukoshi Ltd.
Tokyo, Japan
1987

AD/Tohru Ando
D/Michiharu Takahashi
DF/Design Room, Mitsukoshi Ltd.

265
Shopping Avenue
Ueno Park Ave. Shops Assoc.
Tokyo, Japan
1982

AD/Teruhiko Yumura
D/Tetsuo Fujiwara
I/Teruhiko Yumura
DF/Flamingo Studio Inc., Stereo Studio Inc.

261

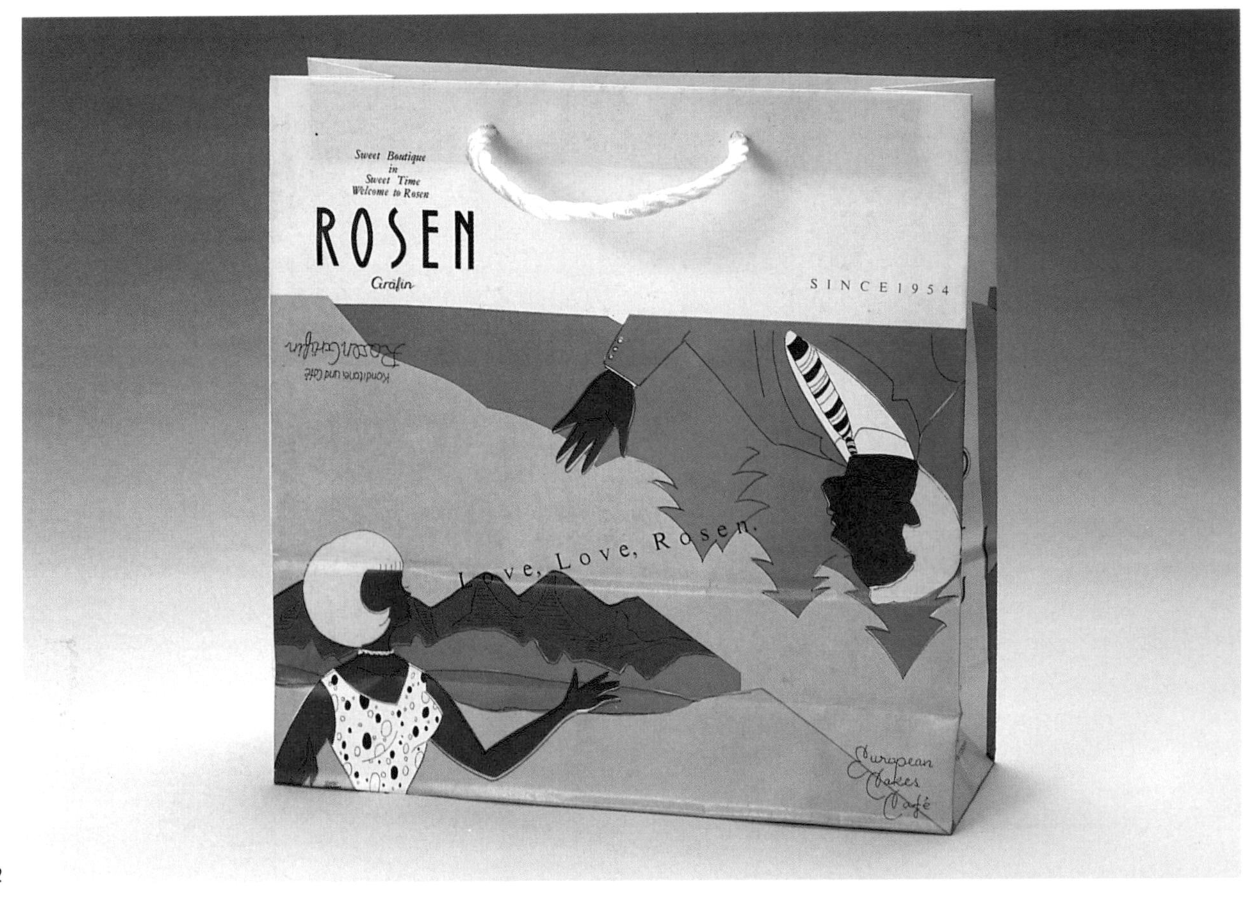

262

'87
British
Fair
MITSUKOSHI

263

MITSUKOSHI
MITSUKOSHI

264

WELCOME TO UENO PARK AVENUE

WELCOME TO UENO PARK AVENUE

265

a

266

b

267

266
スーパーマーケットのオープニング・キャンペーン
西友(春日井西武店)
愛知県春日井市
1977

D／浅葉克己
I／横山 明(a)
P／坂田栄一郎(b)

267
専門店街の紙袋マガジン"ビッグ・レモン"
サンシャインシティ
東京
1987—1988

D／素月道生
DF／エディット・ハウス
I／一噌万佐留(1987年2月号),
村井 茜(1987年7月号),
上原 徹(1987年8月号),
松下 進(1987年10月号),
森 美恵子(1988年1月号),
新野 啓(1988年2月号)

268

269

268
デパートの販促キャンペーン
岩田屋
福岡市
1986

AD／仲畑貴志
D／稲葉欣司
P／坂田栄一郎
DF／岩田屋販促部，仲畑広告制作所

269
ファッション・ショップ・グループ
紙・ビニール
リーザー（FIF グループ）
オランダ・アムステルダム
1983—1987

D／マリアンヌ・ボス
P／ヴィンセンツォ・トラキリオ，キース・ハーゲマン
DF／ザメンヴェルケンデ・オントウェルパース
FIF と呼ばれる，オランダ，ドイツのファッション・ショップ・グループのショッピングバッグ・シリーズ。リーザーは FIF のメンバーである。同社は 1 年に 2 点の割合でバッグを制作しており，それらは冬，あるいは夏のカタログに見ることができる。季節によってデザインが変わるバッグに加えて，1 年中使える標準仕様のバッグがすべてのショップに常時そろっている。

266
Super Market Opening Campaign
The Seiyu, Ltd. (Kasugai Seibu Store)
Aichi Pref, Japan
1977

D／Katsumi Asaba
I／Akira Yokoyama (a)
P／Eiichiro Sakata (b)

267
Boutique Shopping Plaza Paper Bag Magazine “Big Lemon”
Sunshine City Corporation
Tokyo, Japan
1987—1988

D／Michio Suzuki
DF／Edit House Co., Ltd.
I／Masaru Isso (Feb. 1987), Akane Murai (Jul. 1987), Tetsu Uehara (Aug. 1987), Susumu Matsushita (Oct. 1987), Mieko Mori (Jan. 1988), Hiraku Shinno (Feb. 1988)

268
Department Store Sales Campaign
Iwataya Department Stores Co., Ltd.
Fukuoka City, Japan
1986

AD／Takashi Nakahata
D／Kinji Inaba
P／Eiichiro Sakata
DF／Iwataya Department Stores Co., Ltd., Nakahata Co., Ltd.

269
Fashion Shop Group
Paper, Plastic
Leeser (FIF Group)
Amsterdam, The Netherlands
1983—1987

D／Marianne Vos
P／Vicenzo Tracquilio, Kees Hageman
DF／Samenwerkende Ontwerpers
A series of shopping bag for a group of Dutch & German fashion shops called FIF (First in Fashion). Leeser is a member of the FIF Group. They manufacture two different bags a year, each belonging to a winter or summer catalogue. Besides the seasonal bags every shop also has a standard bag which is used all year round.

270
古新聞用紙袋
琉球新報社
那覇
1986

AD／久留米 裕
D／本田良治
I／久留米 裕
DF／トライ
沖縄県の新聞社の古新聞を入れるための紙袋で，家庭の茶の間などに置かれるもの。“沖縄の仲間たち”というタイトルで沖縄の自然を表現している。

271
市販用ショッピングバッグ
ビニール
トライ
東京
1987

AD／久留米 裕
D／本田良治
I／久留米 裕
DF／トライ

270
Paper Bags for Holding Old Newspapers
The Ryuku Shimpo Co., Ltd.
Naha City, Japan
1986

AD／Yutaka Kurume
D／Yoshiharu Honda
I／Yutaka Kurume
DF／Try Co., Ltd.
Paper bag for holding old newspapers, suitable for use inside the home. Bearing the title “Okinawa friends”, the bag portrays the natural beauty of the Okinawa area.

271
Shopping Bag
Plastic
Try Co., Ltd.
Tokyo, Japan
1987

AD／Yutaka Kurume
D／Yoshiharu Honda
I／Yutaka Kurume
DF／Try Co., Ltd.

270 a

270 b

271

273
Stuffed Toy Shop
Paper, Plastic
Another One Co., Ltd.
Tokyo, Japan
1985

D/Akito Honda
DF/Another One Co., Ltd.

274
Bag for LP Records
Plastic
Dusty Miller Co., Ltd.
Tokyo, Japan
1986

I/Osamu Harada
DF/Koji Hompo Co., Ltd.

275
Variety Shop
Plastic
Gendai-Igaku Co., Ltd. (The Show Cruise)
Tokyo, Japan
1988

AD/Haruya Kawamura
D/George Tokoro, Akira Kobayashi
DF/Gendai-Igaku Co., Ltd.

276
Toy Shop
Plastic
Atelier Niki Tiki Co., Ltd.
Tokyo, Japan
1985

D/Toshiko Nishikawa

277
Apparel Boutique
Plastic
Can Co., Ltd.
Tokyo, Japan
1982

D/Hiroshi Fujii
I/Naoko Yoshida
DF/Can Co., Ltd.

278
Boutique
Paper, Plastic
Peppermint Co., Ltd. (Beat Pops By Jemmy's)
Tokyo, Japan
1987

AD/Hiroshi Hattori
D/Kazutoshi Fukawa
DF/Peppermint Co., Ltd.

276

275

277

278

279
ショッピングセンター
ストゥーナー・センター
ノルウェー，オスロー

D／ブルーノ・オルダーニ

280
バラエティ・ショップの'86ハロウィン用ショッピングバッグ
ソニー・プラザ
東京
1986

D／渡辺通正
ハロウィンのテーマ・カラーである黒と黄色を使用。また，ハロウィンの象徴として一般的な"ジャック・オ・ランターン"をあえて避け，キャラクターとして魔女を使って特色を出している。

281
デパートのクリスマスセール
ニーマン・マーカス
アメリカ，テキサス州ダラス
1985

D／アイヴァン・シャマイエフ
DF／シャマイエフ・アンド・ガイスマー・アソシエイツ

279
Shopping Center
Stouner Center
Oslo, Norway

D／Bruno Oldani

280
Variety Shop
'86 Halloween Bag
Sony Plaza Co., Ltd.
Tokyo, Japan
1986

D／Michimasa Watanabe
Designed in black and yellow, the traditional colors for Halloween. The usual Halloween symbol, the "Jack o' Lantern" has been avoided in favor of using the character of a witch in special colors.

281
Department Store Christmas Sale
Neiman Marcus
Dallas, Tex., USA
1985

D／Ivan Chermayeff
DF／Chermayeff & Geismar Assoc.

279

280

281

282
ミュージアム・ショップ
ビニール
ニューヨーク近代美術館
アメリカ, ニューヨーク
1984

AD／五十嵐威暢
D／早瀬和宏, 笹子行美
DF／イガラシステュディオ

283
コンピュータ会社
日本IBM
東京
1987

D／太田雄二
DF／サン・アド

284
魔法びんの販売店用バッグ
象印マホービン
大阪
1987

AD／秋山龍洋, 沢 正一
D／沢 正一
I／舟橋全二
DF／マグナ

283

282
Museum Shop
Plastic
The Museum of Modern Art, New York
New York, USA
1984

AD／Takenobu Igarashi
D／Kazuhiro Hayase, Yukimi Sasago
DF／Igarashi Studio

283
Computer Company
IBM Japan
Tokyo, Japan
1987

D／Yuji Ota
DF／Sun-ad Co., Ltd.

284
Thermos Bottle Retailer
Shopping Bag
Zojirushi Corporation
Osaka, Japan
1987

AD／Tatsuhiro Akiyama, Masakazu Sawa
D／Masakazu Sawa
I／Zenji Funabashi
DF／Magna Inc.

285
Department Store '86 Summer Gift Sale
Matsuzakaya Co., Ltd. (Ueno Store)
Tokyo, Japan
1986

D／Yoshihisa Hirano
DF／Matsuzakaya Co., Ltd. (Ueno Store)
A carrier bag for the summer gift sales. Since it was in the midst of the constellation and comet boom, the Swan (Cygnus), a summer constellation, was chosen as the motif of the design, to present a sense of astronomical romanticism together with cool night breezes.

286
Department Store '86 Cherry Festival Campaign
Matsuzakaya Co., Ltd. (Ueno Store)
Tokyo, Japan
1986

D／Kazuyoshi Katsumata
DF／Matsuzakaya Co., Ltd. (Ueno Store)

284

285
デパートの'86サマーギフトセール
松坂屋(上野店)
東京
1986

D／平野嘉久
DF／松坂屋(上野店)
サマーギフトセールのためのキャリーバッグ。ハレー彗星ブームの渦中にあったため，天体を題材に求め，夏の星座白鳥座をモチーフに，さわやかな天体ロマンの表現を企図している。

286
デパートの'86さくらまつりキャンペーン
松坂屋(上野店)
東京
1986

D／勝俣和良
DF／松坂屋(上野店)

285

286

287

288

287
ファッション会社
ビニール
エスプリ・デ・コーポレーション
アメリカ，サンフランシスコ
1987

D／マイケル・ヴァンダビル
P／フィル・ハント・ステュディオス
DF／ヴァダンビル・デザイン
良質のコットンを使ったリネン，タオル，アクセサリー類を含むエスプリの最新ブランド，エスプリ・バス＆ベッドのためにデザインされたショッピングバッグ。

288
洋菓子店
Sagan
小田原市
1986

D／中谷匡児
バッグに見えるロゴタイプは，フランス的な上品さを意図してデザインされたもの。包装紙のデザインはロゴタイプのパーツを分解し，再構成してつくられた。

289
デパート
東武デパート
東京
1987

AD／高橋新三
D／高橋新三，川崎修司
DF／ヘッズ
A／マッキャンエリクソン博報堂

290
菓子店
紙，ビニール
ハタダ
松山市
1985

D／イングデザイン研究所

291
雑貨店
ラフェーテ
大阪
1987

D／鳥山大樹
DF／バード・デザインハウス

287
Fashion Company
Plastic
Esprit De Corporation
1987

D/Michael Vanderbyl
P/Phil Hunt Studios
DF/Vanderbyl Design
This shopping bag was designed for Esprit Bath & Bed, a new line of Esprit products featuring fine cotton bed linens, towels and accesories.

288
Confectionery
Sagan
Odawara City, Japan
1986

D/Kyoji Nakatani
The logo on the bag is designed to add a touch of French elegance. The design of the wrapping paper is created by reassembling the parts of the logo.

289
Department Store
Tobu Department Stores Co., Ltd.
Tokyo, Japan
1987

AD/Shinzo Takahashi
D/Shinzo Takahashi, Shuji Kawasaki
DF/Heads Inc.
A/McCann-Erickson Hakuhodo Inc.

290
Confectionery
Paper, Plastic
Hatada Co., Ltd.
Matsuyama City, Japan
1985

D/…ing Design Institute

291
Grocery Store
Laféte Co., Ltd.
Osaka, Japan
1987

D/Taiki Toriyama
DF/Bird Design House

289

290

291

292
レストラン，料亭
花外楼
大阪
1983

AD／吉羽敏郎
D／高田雄吉
DF／アイ・エフ・プランニング
レストラン"アイル・モレ"のテイクアウトはパンが中心なので，シンプル＆ローコストが最優先された。背景はテクスチャーのある和紙を写真撮影したもの。水の流れを表現したパターンを白抜きでクラフト紙に印刷した。"アイル・モレ"はインドネシア語で"美しい水"の意味。もう1点は親会社の高級料亭・花外楼のテイクアウト食品用のもの。

293
せんべい屋
ねぼけ堂
熊本県八代市
1987

D／イングデザイン研究所
おかき，あられ，せんべいのイメージを小柄パターンで表わし，その大きな分割で"味のモダン"を表現しようとしている。

292

293

292
Restaurant
Kagairoh Co., Ltd.
Osaka, Japan
1983

AD/Toshiro Yoshiba
D/Yukichi Takada
DF/I.F.Planning Co., Ltd.
The highest priority in this design is given to simplicity and low cost, the major takeout items the restaurant "Air Molék" being bread. The background is a photo of traditional Japanese paper which has a textured surface. The pattern is reversed to white to present the flow of water on kraft paper. "Air Molék" is an Indonesian phrase that means "beautiful water". The other design here is for takeout food from the first class cuisine "Kagairoh", the parent company of Air Molék.

293
Rice Cookies Shop
Nebokedo
Kumamoto Pref, Japan
1987

D/…ing Design Institute
Small individual patterns on this work present images of Japanese rice cookies okaki, arare and sembei, while a large partition in the design presents the "modern taste" of the products.

294

294
宝石店
ラリー・ジュエリー
ホンコン
1987

AD/カン・タイ-クン
D/ジョンソン・ヒュアン, エディ・ユー
I/ジョンソン・ヒュアン
DF/SS デザイン・アンド・プロダクション

295
衣料品店
プリンシペ
イタリア, フィレンツェ
1978

D/レオナルド・バグリオーニ
DF/ストゥディオ・レオナルド・バグリオーニ

294
Jewelry Store
Larry Jewelry
Hong Kong
1987

AD/Kan Tai-keun
D/Johnson Huang, Eddy Yu
I/Johnson Huang
DF/SS Design & Production

295
Clothing Store
Principe Spa.
Firenze, Italy
1978

D/Lenardo Baglioni
DF/Studio Leonardo Baglioni

295

296

297

296
スクリーン製造会社
大日本スクリーン製造
東京
1982

D／永井一正

297
"グレース・ブラザーズ"
グレース・ブラザーズ・パーティー・リミテッド
オーストラリア, シドニー
1987

D／ケン・ケイトー
DF／ケイトー・デザイン・インコーポレイテッド

296
Screen Manufacturing Company
Dainippon Screen Manufacturing Co., Ltd.
Tokyo, Japan
1982

D／Kazumasa Nagai

297
"Grace Bros."
Grace Bros. Pty. Ltd.
Sydney, Australia
1987

D／Ken Cato
DF／Cato Design Inc.

298

299

298
デパートのクリスマスセール
京王百貨店
東京
1987

AD／石井慎悟
D／合原久美
Producer／相原英之
DF／エージー

299
洋菓子・喫茶店
ジーセンコンフェクト
神戸市
1970

D／早川良雄

298
Department Store Christmas Sale
Keio Department Store Co., Ltd.
Tokyo, Japan
1987

AD／Shingo Ishii
D／Kumi Gohara
Producer／Hideyuki Aihara
DF／A & G Co., Ltd.

299
Coffee & Confectionery Shop
G-sen Confect Co., Ltd.
Kobe City, Japan
1970

D／Yoshio Hayakawa

300

301

300
デパートのクリスマスセール
松坂屋
東京
1987

D／斎藤智久
DF／松坂屋
北欧調のイラストでまとめ，グレー地を生かしたシックな色調の中にクリスマスをさりげなく表現。

301
クリスマス用のショッピングバッグ
スーパーバッグ
東京
1986

AD／スーパーバッグ S.P.C. 室
D／石崎洋之
DF／ヴァーナル・アンド・カンパニー

300
Department Store Christmas Sale
Matsuzakaya Co., Ltd.
Tokyo, Japan
1987

D／Tomohisa Saito
DF／Matsuzakaya Co., Ltd.
In an illustration with a Scandinavian touch, a chic color utilizes the gray background to produce a light-hearted Christmas mood.

301
Shopping Bag for Christmas
Super Bag Co., Ltd.
Tokyo, Japan
1986

AD／S.P.C.Room, Super Bag Co., Ltd.
D／Hiroyuki Ishizaki
DF／Vernal And Company Ltd.

302
果物店
銀座千疋屋
東京
1975

D／斎藤義雄

302
Fruit Shop
Ginza Senbikiya
Tokyo, Japan
1975

D／Yoshio Saito

302

303
雑貨店
ドノバン(モノショップ・フィール)
松山市
1987

AD／池田武仁
D／池田武仁, 光盛香奈子
DF／ブジンアート
品物の味わいや特色を生かすのは使う人自身という考えから，販促物・包装紙など，すべてモノトーンが基調になっている。

303
Grocery Store
Donoban Inc. (Mono-shop "Feel")
Matsuyama City, Japan
1987

AD／Takehito Ikeda
D／Takehito Ikeda, Kanako Mitsumori
DF／Buzin Art
Sales items, wrapping paper, etc. all have acquired monotone hues in response to the taste of the purchaser, whose preferences come to dictate the particular colors and qualities of the product.

303

304

305

306

307

304
印刷会社
寿精版印刷
大阪
1988

AD／五十嵐威暢
D／森田穂波
DF／イガラシステュディオ

305
書籍用ショッピングバッグ
主婦の友社
東京
1982

D／藤井嘉彦
DF／シティ・ユニオン

306・307
婦人靴店
ヨシノヤ
大阪
1984

D／鳥山大樹
DF／バード・デザインハウス
アフターユーズも考慮して，「どの年代の人にも持ってもらえるペーパーバッグ」が制作意図。包装紙とアンサンブルのデザイン。

304
Printing Company
Kotobuki Seihan Printing Co., Ltd.
Osaka, Japan
1988

AD／Takenobu Igarashi
D／Honami Morita
DF／Igarashi Studio

305
Shopping Bag for Books
Shufunotomo-sha Co., Ltd.
Tokyo, Japan
1982

D／Yoshihiko Fujii
DF／City Union Co., Ltd.

306・307
Ladies' Shoe Shop
Yoshinoya
Osaka, Japan
1984

D／Taiki Toriyama
DF／Bird Design House
Intending to design "a paper bag for every generation", this design even considers after-use, and coordinates with the wrapping paper.

308
キャラクター商品用のショッピングバッグ
サントリー
大阪
1986

AD／山登道雄
D／牛島志津子
DF／サントリー・デザイン部

309
宅急便
大和運輸
東京
1986

AD／石崎洋之
D／田原 香
DF／ヴァーナル・アンド・カンパニー
A／スーパーバッグ S.P.C. 室

310
"Intel '85"
ビニール
バッサーニ・ティチーノ
イタリア, ミラノ
1985

D／ハインツ・ウァイブル
DF／シグノ SRL

308
Shopping Bag for Character Products
Suntory Limited
Osaka, Japan
1986

AD／Michio Yamato
D／Shizuko Ushijima
DF／Suntory Limited

309
Home Delivery Service
Yamato Un'yu Co., Ltd.
Tokyo, Japan
1986

AD／Hiroyuki Ishizaki
D／Kaori Tahara
DF／Vernal And Company Ltd.
A／S.P.C.Room, Super Bag Co., Ltd.

310
"Intel '85"
Plastic
Bassani Ticino
Milano, Italy
1985

D／Heinz Waibl
DF／Signo SRL

308

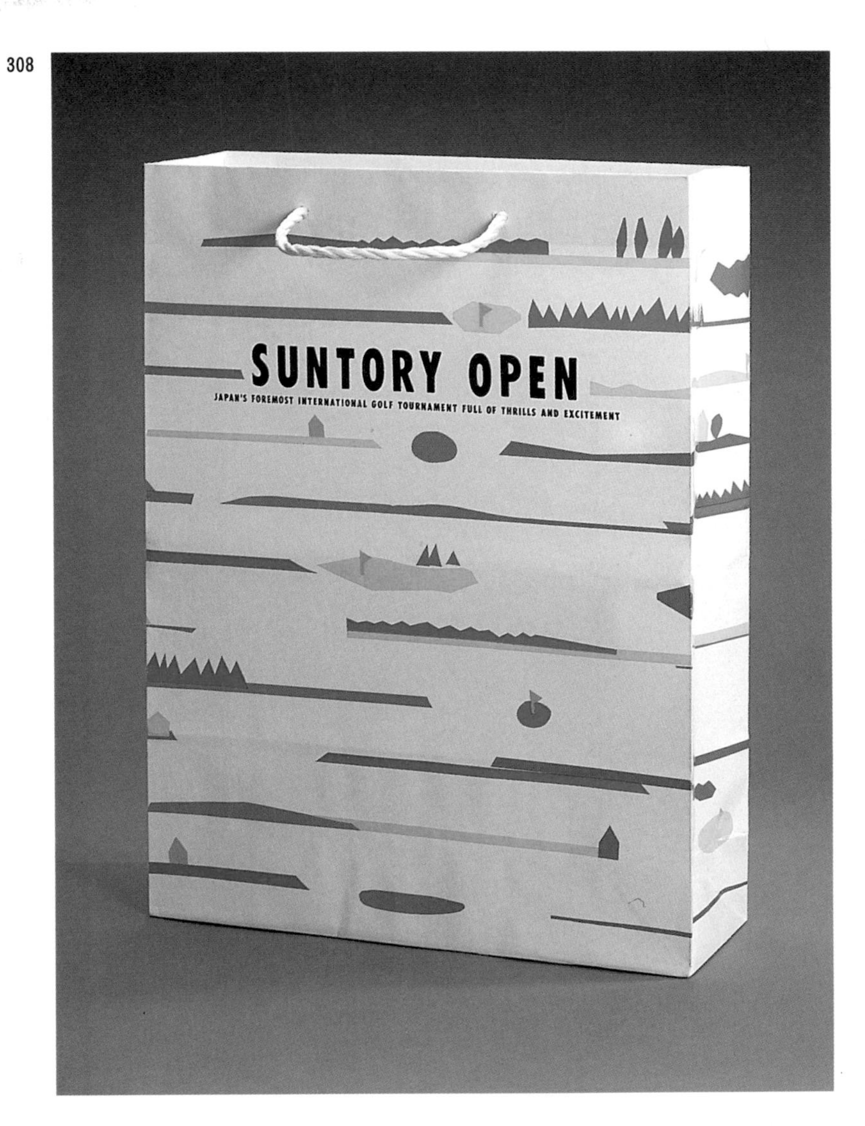

309

310

311

311
ブーツ用ギフトバッグ
ザ・ブーツ・カンパニー
イギリス, ノッチンガム
1987

AD／リン・トリケット, ブライアン・ウェッブ
D／リン・トリケット, ブライアン・ウェッブ, フィオナ・スケルシー
DF／トリケット・アンド・ウェッブ・リミテッド

311
Gift Bag for Boots
The Boots Company
Nottingham, England
1987

AD／Lynn Trickett, Brian Webb
D／Lynn Trickett, Brian Webb, Fiona Skelsey
DF／Trickett & Webb Ltd.

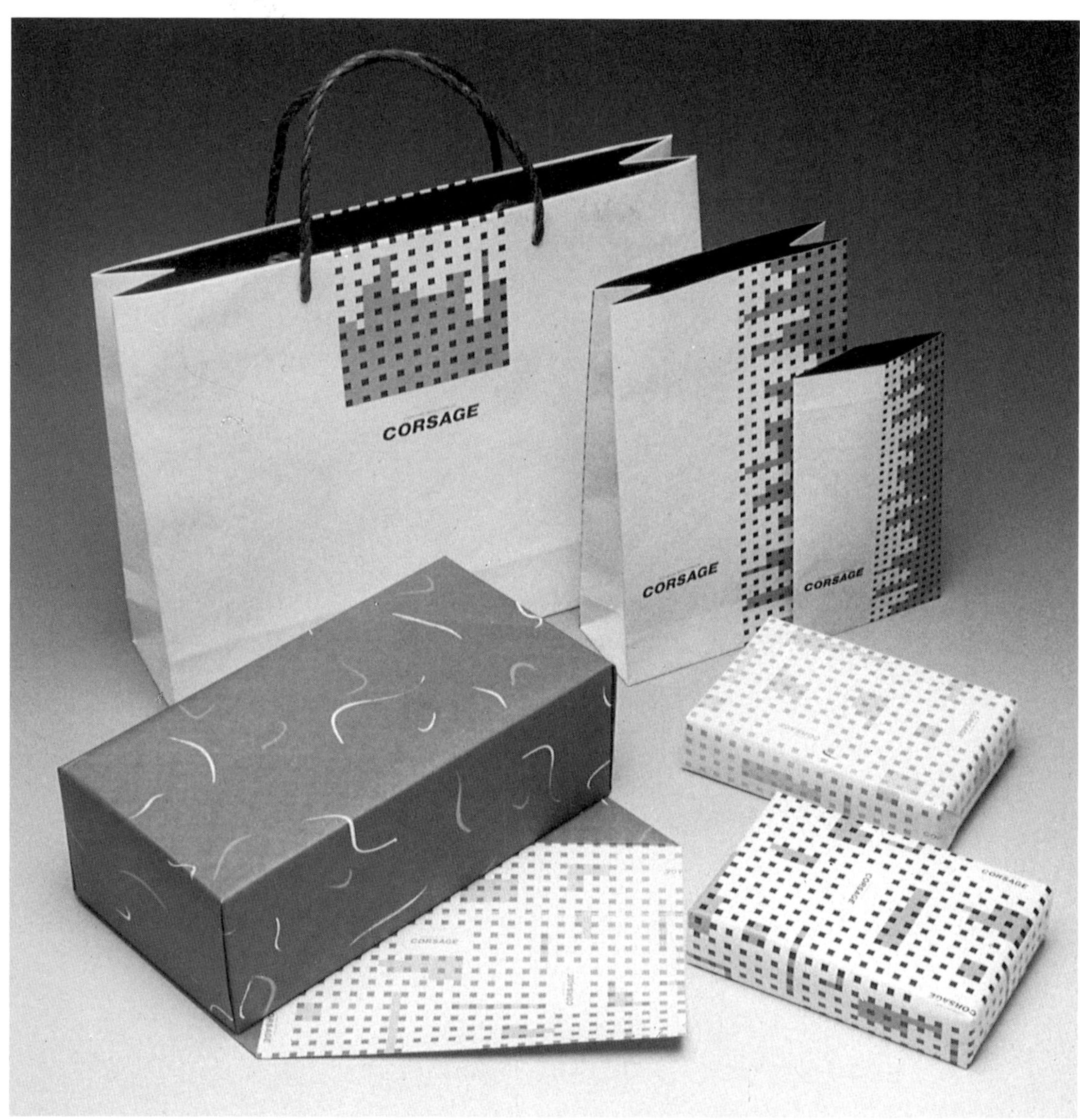

312

313

312
化粧品店
コサージュ
熊本市
1986

D／イングデザイン研究所
店名のもつ柔らかい響きとは反対に，デザインは男っぽく，マンハッタンのシルエットでクールなヴィジュアル効果を意図している。

313
香水会社
ブッチ
イタリア，ベルガーノ
1984

D／ハインツ・ヴァイブル
DF／シグノ SRL

312
Cosmetics Store
Corsage
Kumamoto City, Japan
1986

D／…ing Design Institute
With the introduction of a silhouette of Manhattan in contrast to the soft mood produced by the store's name is contrasted against a cool, masculine effect.

313
Perfume Company
Bucci
Bergamo, Italy
1984

D／Heinz Waibl
DF／Signo SRL

a

b

c

d

314

314
和菓子
赤坂青野
東京
1987

AD／村越 襄
D／村越 襄, 鈴木 薫
Artist／加山又造
DF／村越襄デザイン室
A／アンゼン・パックス
和菓子の老舗，赤坂青野の本店ビル落成を記念して一新されたパッケージング・システムの一部。春(b), 夏, 秋(c), 冬(d)用の 4 種とレギュラー, 祝儀, 無祝儀(a)用がラインナップされている。

315
おかき店
弥乃一
東京
1987

AD／市口清一
D／市口清一, 沢田真輝, 堀 久夫
DF／オーエムシー

316
銘菓"管絃祭"
広島民芸社
広島市
1980

D／藤重日生

314
Japanese Style Confectionery Shop
Akasaka-Aono
Tokyo, Japan
1987

AD／Jo Murakoshi
D／Jo Murakoshi, Kaoru Suzuki
Artist／Matazo Kayama
DF／Murokoshi Jo Design Shitsu
A／Anzen Packs
Part of the packaging system of Akasaka-Aono, an old-established Japanese style confectionery, was renewed in commemoration of the opening of the new building. The packaging system consists of bags for the four seasons, regular bags, congratulatory bags, non-congratulatory bags, etc.

315
Rice Cookies Shop
Yanoichi
Tokyo, Japan
1987

AD／Seiichi Ichiguchi
D／Seiichi Ichiguchi, Maki Sawada, Hisao Hori
DF／OMC, Inc.

316
Japanese Confection Brand "Kangensai"
Hiroshima-mingei-sha Co., Ltd.
Hiroshima City, Japan
1980

D／Teruo Fujishige

315

316

317
日本酒
宮坂醸造
長野市
1987

AD／石崎洋之
D／田原 香
DF／ヴァーナル・アンド・カンパニー
A／スーパーバッグ S.P.C. 室

318
レジャー開発地区
日本レジャー開発
栃木県那須
1987

AD／八尾武郎
D／村上和彦
DF／YAO デザイン研究所

319
レストラン
椿
東京
1982

D／片岡 脩
DF／シウ・グラフィカ

320
民芸品店
日本通運遠野営業所
岩手県遠野市
1985

D／斎藤喜郎
DF／ハイプロダクション
A／博報堂盛岡

321
みやげもの店
御土産処えいと
新潟市湯沢温泉
1986

D／白川こう吉, 大谷雅子
I／白川こう吉
A／立見カートン
湯沢温泉のみやげもの店のショッピングバッグ。店名の"えいと"を湯沢の"八"で表現しているが, これは同時に"山"も示している。

317
Carrier Bag for Sake Bottle
Miyasaka Jyozo Co., Ltd.
Nagano City, Japan
1987

AD／Hiroyuki Ishizaki
D／Kaori Tahara
DF／Vernal And Company Ltd.
A／S.P.C.Room, Super Bag Co., Ltd.

318
Bag for Leisure Development District
Nippon Leisure Kaihatsu Co., Ltd.
Tochigi Pref, Japan
1987

AD／Takeo Yao
D／Kazuhiko Murakami
DF／Yao Design Institute Inc.

317

318

319
Restaurant
Tsubaki
Tokyo, Japan
1982

D/Shu Kataoka
DF/Siu Graphica Inc.

320
Folkcraft Shop
Nippon Express Co., Ltd.
Tono City, Iwate Pref, Japan
1985

D/Kichiro Saito
DF/Hi-Production Co., Ltd.
A/Hakuhodo, Inc.

321
Souvenir Shop
Souvenir Shop "Eight"
Yuzawa Hotspring Resort, Japan
1986

D/Kokichi Shirakawa, Masako Otani
I/Kokichi Shirakawa
A/Tachimi Carton Co., Ltd.
The shopping bag of a souvenir shop in Yuzawa Hotspring Resort. The Kanji character meaning "8" is used in the design stands, standing not only for the shop's name "Eight" but for the shape of the mountains around Yuzawa.

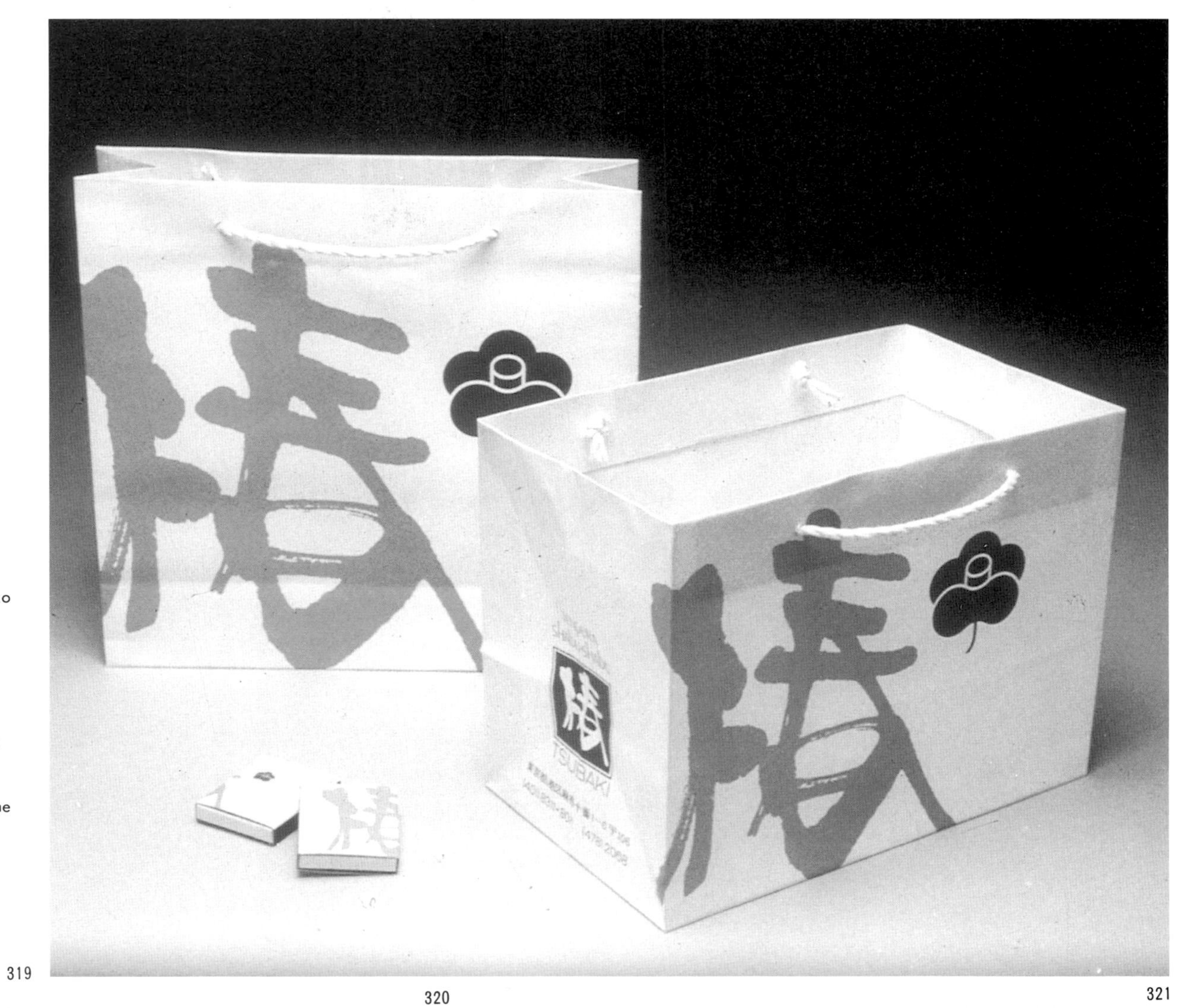

319

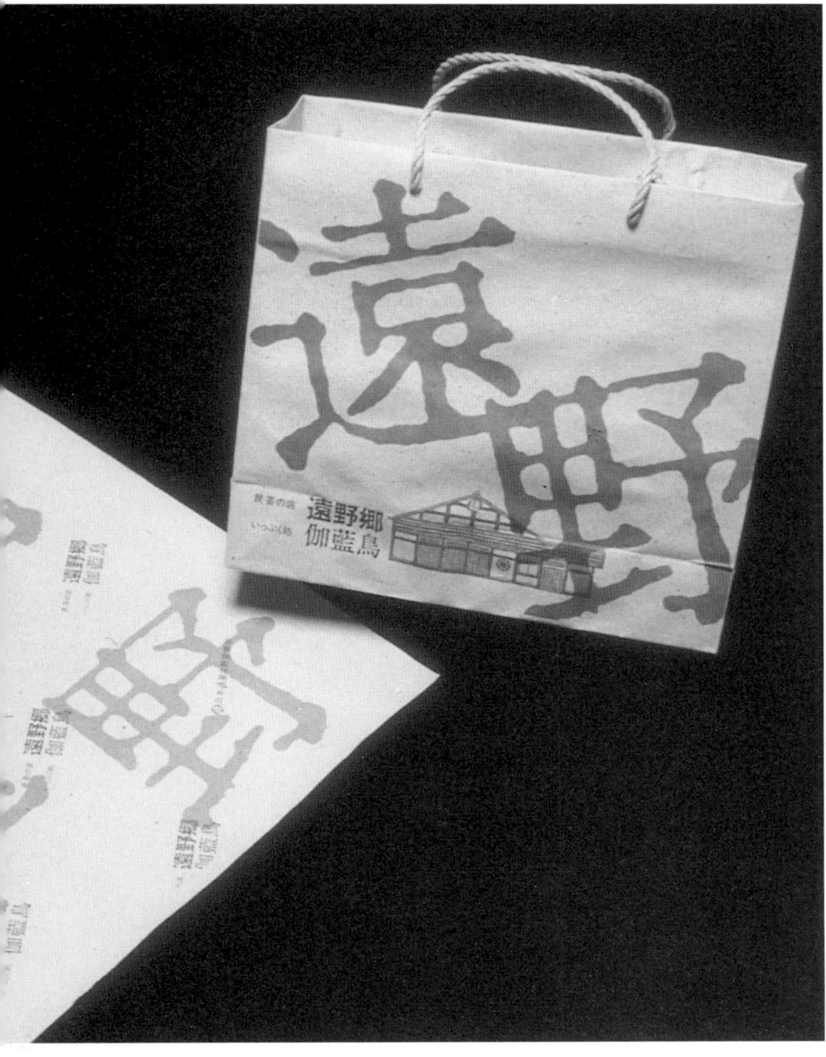

320

321

322
煮込みうどん店
山本屋本店
名古屋
1986

D／高木秀道
DF／ルーム・ワン
ざっくりとした筆のタッチを取り入れたデザイン。ラベルシールには商品の特性が記されている。

322
Japanese Noodle Shop
Yamamotoya Honten
Nagoya City, Japan
1986

D／Hidemichi Takagi
DF／Design Office Room-One
A simple design introducing the crisp touch of a writing brush. A label bears the information on the features of the product.

323
個展のためのショッピングバッグ
東京デザイナーズスペース
東京
1987

D／川崎修司, 高橋新三
P／能津喜代房
DF／ヘッズ
東京デザイナーズスペースでの個展のために，取引のある10社を選んで自由に制作したもの。

323
Shopping Bag Works for the One-man Exhibit
Tokyo Designer's Space
Tokyo, Japan
1987

D／Shuji Kawasaki, Shinzo Takahashi
P／Kiyofusa Nozu
DF／Heads Inc.
Optionally designed works on selected items for ten clients for one-man exhibiting at the Tokyo Designer's Space.

322

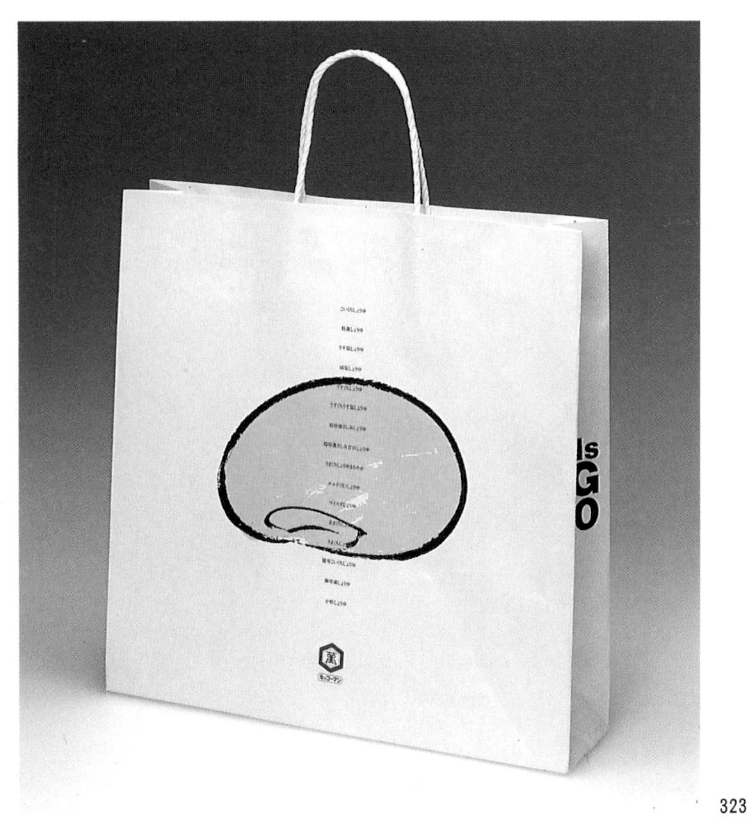

323

324
和菓子屋
嵯峨野松風
京都
1986

AD／岩崎堅司
D／片岡　脩

324
Japanese Style Confectionery Shop
Sagano Matsukaze Co., Ltd.
kyoto, Japan
1986

AD／Kenji Iwasaki
D／Shu Kataoka

324

325
レストラン
柿安本店
三重県桑名市
1980

D／柿安本店企画開発課

326
観光みやげもの売場
嬬恋村観光物産案内所
群馬県嬬恋村
1986

I／野村たかあき
DF／でくの房
嬬恋村の観光みやげもの売場のショッピングバッグで，創作民話をキャラクター化してデザインした。

325
Restaurant
Kakiyasu-Honten
Kuwana City, Mie Pref., Japan
1980

D／Planning Section, Kakiyasu-Honten Co., Ltd.

326
Souvenir Shop
Tsumagoi-mura Local Products Center
1986

I／Takaaki Nomura
DF／Dekunobo Studio
This bag, designed for use in souvenir stores, illustrates folklore of the village translated into images.

325

326

327
鮭のショッピングケース
別海漁業協同組合
北海道別海町
1987

D／竹内二朗
DF／ミッキーハウス
A／博報堂札幌
贈答用に開発したものだが，このまま発送できるようになっている。従来のパッケージと異なり，保管場所もとらず，何本かを一度に運ぶこともできる。街中を持って歩けるデザインを意図している。

327
Shopping Case for Salmon
Bekkai Fishery Cooperative Assoc.
Hokkaido, Japan
1987

D／Jiro Takeuchi
DF／Micky House Inc.
A／Sapporo Branch, Hakuhodo, Inc.
A packaging product developed for a gift. It can be shipped for delivery as is, and In addition occupies less space. Several packages can be carried at one time as well, overcoming the disadvantages of conventional bags. This bag is to match a fashionable street.

327

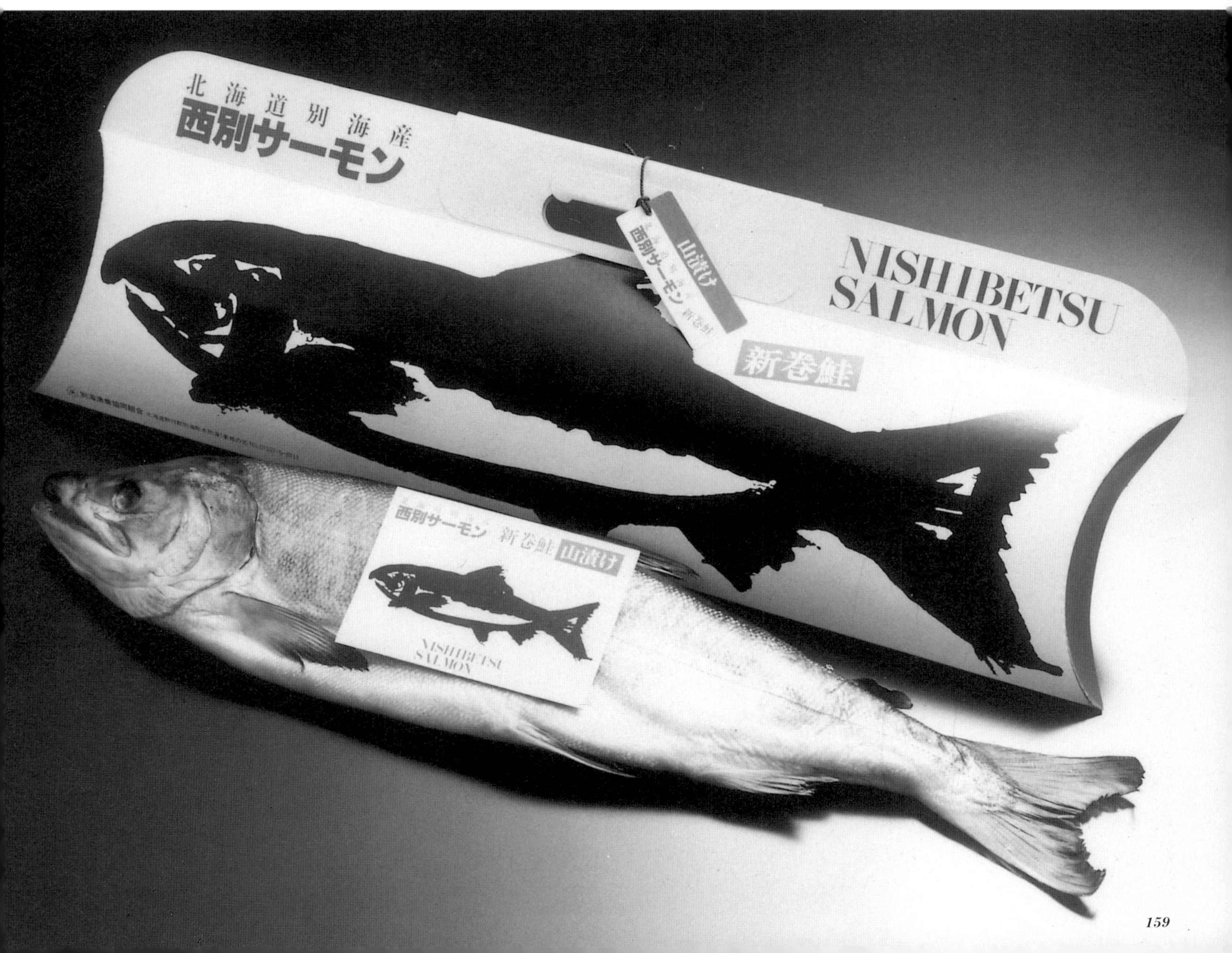

328

328
漬物屋
紙+ワックス
若菜
名古屋
1984

D／矢萩喜従郎
DF／キジュウロウ・ヤハギ

328
Pickle Shop
Paper + Wax
Wakana Co., Ltd.
Nagoya, Japan
1984

D／Kijuro Yahagi
DF／Kijuro Yahagi Co., Ltd.

329

329
桐工芸店
会津桐専門店・やまき
福島県喜多方市
1985

AD／熊谷英博
D／熊谷英博, 磯田日出夫

329
Paulownia Wood Craft Shop
Yamaki Co., Ltd.
Kitakata City, Japan
1985

AD／Hidehiro Kumagai
D／Hidehiro Kumagai, Hideo Isoda

330

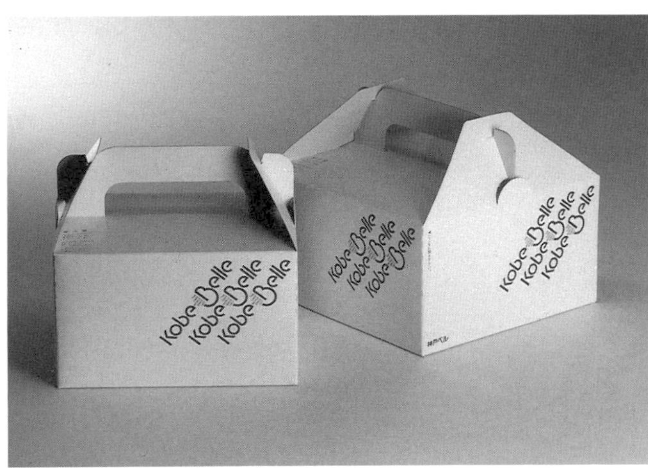

331

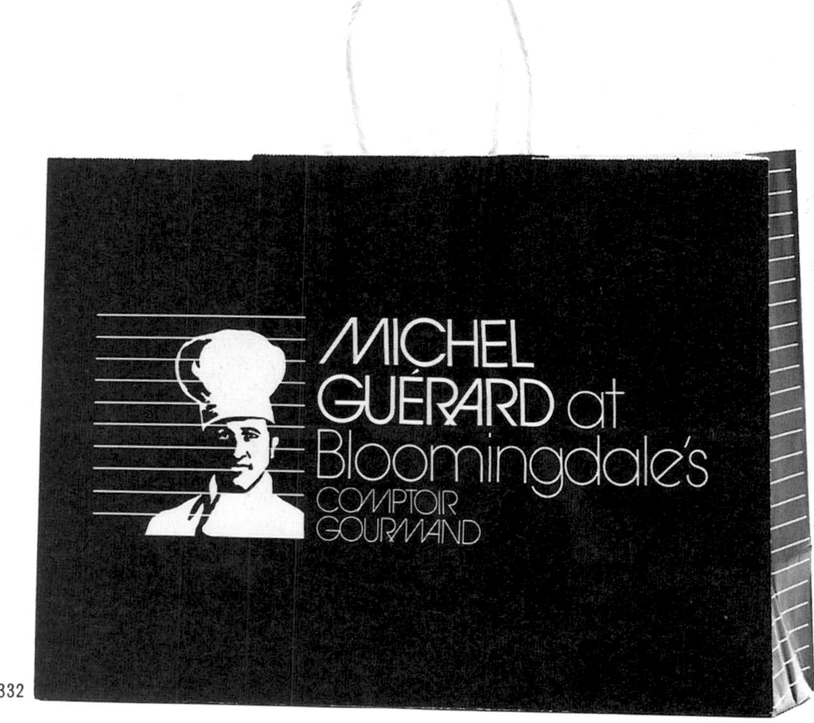

332

330・331
洋菓子会社
ベル(神戸ベル)
神戸市
1983―1985

D／ヘルムート・シュミット
Planner／西尾捷太
A／JMS

332
デパート
ブルーミングデイルズ
アメリカ, ニューヨーク
1982

D／ロバート・ガーシン
DF／ロバート・P・ガーシン・アソシエイツ
ブルーミング・デイルズのフード・ストアー"ミッシェル・ギュラール・コンプター・グールマン"のショッピングバッグ。

330・331
Confectionery Company
Belle Co., Ltd. (Kobe Belle)
Kobe City, Japan
1983―1985

D／Helmut Schmid
Planner／Shota Nishio
A／Japan Marketing Service Co., Ltd.

332
Department Store
Bloomingdale's
New York, USA
1982

D／Robert Gersin
DF／Robert P. Associates, Inc.
This bag was designed for Michel Guerard Comptor Gourmand convenience foods, which were part of the Bloomingdale's specialty food department.

333

333
洋品雑貨店
ポルトフィーノ
大阪
1985

D／高月太郎
DF／ポルトフィーノ

334
ファッション・ブランド"フェンディ"
アオイ
東京
1984

D／フェンディ(イタリア)
DF／下島包装開発

333
Variety Shop
Portofino
Osaka, Japan
1985

D／Taro Takatsuki
DF／Portofino

334
Fashion Brand "Fendi"
Aoi Co., Ltd.
Tokyo, Japan
1984

D／Fendi (Italy)
DF／Shimojima Packaging Co., Ltd.

334

335

335
結婚式
紙, ビニール
ロイヤルパレス平安閣
神戸市
1986

D／佐古田英一
DF／サン・リーズ
A／新光美術
鶴と亀(亀甲)，紅と白という祝事の伝統的な形と色を現代的にヴィジュアル化することを意図した。

335
Wedding Ceremony Hall
Paper, Plastic
Royal Palace Heiankaku
Kobe City, Japan
1986

D／Eiichi Sakota
DF／Sun-Reeds Co., Ltd.
A／Shinko Art Inc.
Seeking to apply a feeling of the modern visualization method to the use of the "crane and turtle back" and "red and white", both of which are the traditional Japanese symbols for celebratory events.

336
ブティック
K. Shop, Takako Kubo
東京
1987

AD／河北秀也
D／明星秀隆

336
Boutique
K.Shop, Takako Kubo
Tokyo, Japan
1987

AD／Hideya Kawakita
D／Hidetaka Myojo

336

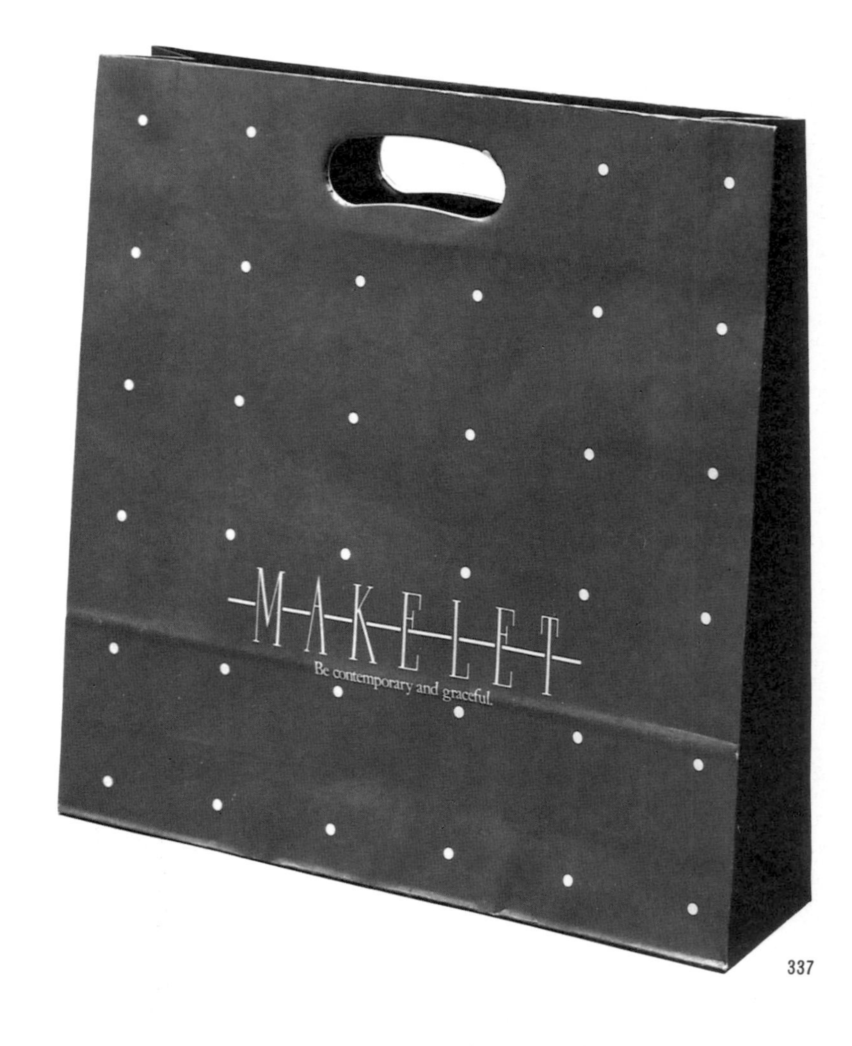

337

337
子供服ブティック
アイドル・メイクレット
東京
1987

D／アイドル・メイクレット

337
Children's Clothing Boutique
Idol Makelet Co., Ltd.
Tokyo, Japan
1987

D／Idol Makelet Co., Ltd.

338

338
ファッション・ブティックのクリスマス・キャンペーン
ザ・ギンザ
東京
1987

D／太田和彦
レギュラーのショッピングバッグのデザインをベースに制作されたクリスマス・キャンペーン用のショッピングバッグ。レギュラーのショッピングバッグの色をクリスマスカラーのグリーン，レッドに変えて，ザ・ギンザ・マークに雪をのせたデザイン。

338
Fashion Boutique Christmas Campaign
Shiseido Boutique The Ginza
Tokyo, Japan
1987

D／Kazuhiko Ota
A basic shopping bag design has been adapted for use during the Christmas season. Snow falls on The Ginza identification mark, using the Christmas colors of red and green.

339

339
グルメ・フード・ショップ
オークヴィル・グロッサリー
アメリカ, カリフォルニア州ナパ・バレー
1980

D/マイケル・マナリング
DF/オフィス・オブ・マイケル・マナリング

339
Gourmet Foods Retailer
Oakville Grocery Co.
Napa Valley, CA, USA
1980

D/Michael Manwaring
DF/The Office of Michael Manwaring

340
メンズ・ブティック
Y. Inohana & Sons
京都
1978

D/位野花敏郎

340
Men's Boutique
Y. Inohana & Sons
Kyoto, Japan
1978

D/Toshiro Inohana

340

341

342

341
ブティック
布
ポップ・インターナショナル(USPP)
東京
1986

D/山下朝子
DF/USPP

342
アパレル・ブティック
モーニン・ムー
大阪

D/モーニン・ムー

341
Boutique
Cloth
Pop International Co., Ltd.
(USPP)
Tokyo, Japan
1986

D/Asako Yamashita
DF/United States of Paradise
Park

342
Apparel Boutique
Mornin' Mou
Osaka, Japan

D/Mornin' Mou

343
駄菓子屋
ハニー(ハラッパ A)
東京
1984

AD／佐久間由紀子
D／友枝康二郎
DF／インタレスト
“ハラッパ A”は，駄菓子と昔なつかしいおもちゃを売っている店。ティン・トイのネジを巻くカギの形が店のシンボルマークになっている。

343
Sweets and Toys Shop
Honey Co., Ltd. (Harrapa A)
Tokyo, Japan
1984

AD／Yukiko Sakuma
D／Kojiro Tomoeda
DF／Interest Inc.
“Harrapa A” is a shop selling sweets and toys of yesteryear. The shape of a winding key, a shape which anyone who has a tin toy will recognise, has become the shop's symbol.

343

344

344
ブティック
アン
大阪

D／ann

344
Boutique
ann
Osaka, Japan

D／ann

345

346

345
ショッピングバッグ
アンガス・アンド・クート
オーストラリア，シドニー
1987

AD／レイモンド・ベネット
D／デジリー・フーケヴィーン
DF／レイモンド・ベネット・デザイン・アソシエイツ

346
ビューティ・サロン
エゴ
オーストラリア，クイーズランド

D／バリー・タッカー
I／エリザベス・シュルーズ
DF／バリー・タッカー・デザイン
A／N/A
クイーズランドの一等地，サンクチュアリー・コーブ・リゾートに位置する，ヘア・スタイリングとビューティーサロンを兼ねた店，"エゴ"のショッピングバッグ。

345
Shopping Bag
Angus & Coote Pty. Ltd.
Sydney, Australia
1987

AD／Raymond Bennett
D／Desiree Hoogeveen
DF／Raymond Bennett Design Associates

346
Beauty Salon
Ego Hair Styling And Beauty Therapy
Queensland, Australia
1987

D／Barrie Tucker
I／Elizabeth Schlooz
DF／Barrie Tucker Design Pty. Ltd.
A／N/A
A shopping bag for Ego Ladies and Mens Hair Styling and Beauty Therapy Salon, situated at the prestigious Sanctuary Cove Resort, Queensland.

347
デパート
プランタン銀座
東京
1987

D／プランタン・パリ，諸星宗宏
DF／プランタン銀座宣伝企画部

347
Department Store
Printemps Ginza S.A.
Tokyo, Japan
1987

D／Printemps Paris, Munehiro Morohoshi
DF／Printemps Ginza S.A.

347

348

348
ビデオ編集スタジオ
サージ・スタジオ
神奈川県
1987

CD／海老名 淳
D／竹智 淳
DF／バーヴ
脱都会型プロダクションとして近郊の海浜住宅地区に設立されたビデオ編集スタジオのペーパーバッグで，このスタジオのビジュアル・アンデンティティの一環として制作された。

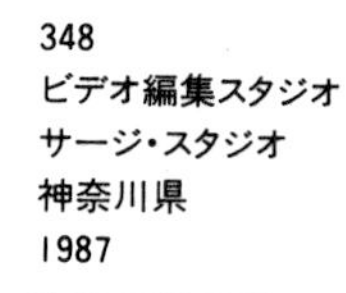

349
ファッション・ブランド
KENZO Paris
東京，パリ

D／KENZO S. A., Concept Group

348
Video Editing Studio
Serge Studio
Kanagawa Pref., Japan
1987

CD／Jun Ebina
D／Jun Takechi
DF／Verve Inc.
A paper bag from a video editing studio which is located in a suburban residential area along the coast, as a production firm of the urban-escape type. The bag is designed as one of the items for visual identity.

349
Fashion Brand
KENZO Paris
Tokyo, Japan;Paris, France

D／KENZO S. A., Concept Group

349

350

351

350
メンズ・ブティック
メンズショップ・マルヤマ(サウス・モルトン)
京都
1987

D／嶋 高宏
ジーンズというカジュアルなアイテムを生かすため，“あたたかいナチュラルな感覚”を重視。高級感と都会的なエレガンスをできる限り広くとった空間で表現している。

351
子供服店
子供服 Nice
名古屋
1981

D／山内瞬葉
奥のバッグはマチが薄くとってあり，手前の大型のバッグはマチが厚くとってある。紙はさらしクラフト。

350
Men's Boutique
Men's Shop Maruyama (South Molton)
kyoto, Japan
1987

D／Takahiro Shima
In order to enhance the chracteristics of the casual item —Jeans— importance is put on the "warm and natural touch" for paper material, logos and markings. Designed to express the item in a space filled with a maximum of urban elegance and prestige.

351
Children's Clothing Shop
Nice
Nagoya City, Japan
1981

D／Shunyo Yamauchi
The bag at the back has a thin gore while the bag at front has thick gore. Bleached kraft paper is used.

352
食品ブランド"豆の旅"
フジッコ
兵庫県西宮市
1985

AD／辻本有邦
D／辻本有邦，板倉悦次
DF／サン・デザイン・アソシエーツ
ショッピングバッグだが蛍光看板的な効果も意図している。遠方からもシンボルが印象的なように大胆に表現。特に面白さを強調し，食品会社の清潔感がポイントになっている。

353
バッテリー・フラッシュライト・メーカー
ビニール
クロアチア
ユーゴスラヴィア，ザグレブ
1987

D／ボリス・リュービチック
P／ピーター・ダバック
DF／スタジオ・インターナショナル
"クロアチア"はユーゴスラヴィア最大のバッテリー・フラッシュライト・メーカー。80周年記念のヴィジュアル・アイデンティティは，三角錐と立方体と球体からなる数字の"80"で構成されている。このビニールバッグはそうした販促用ツールのうちのひとつ。

352
Food Brand "Mame-no-tabi"
Fujikko Co., Ltd.
Nishinomiya City, Japan
1985

AD／Arikuni Tsujimoto
D／Arikuni Tsujimoto, Etsuji Itakura
DF／Sun Design Associates
This is a shopping bag, but it aims in parallel at the effect of a fluorescent sign board. A bold design is adopted so that the symbol may be impressive even from a distance. Not only the unique character of the product, but also a clean snd hygienic image for the food manufacturing company, is emphasized.

353
Battery & Flashlight Manufacturer
Plastic
Croatia
Zagreb, Yugoslavia
1987

D／Boris Liubičić
P／Petar Dabac
DF／Studio International
"Croatia" is, the biggest Yugoslav manufacturer of batteries and flashlights. The visual identity for the 80th anniversary centers on the numeral 80, constructed from a pyramid, a cube, and a sphere. This plastic bag is one in a series of promotional articles.

352

353

354
ファッション・ブティック
ザ・ギンザ
東京
1985

D／仲條正義

355
音響機器会社
ケンウッド
東京
1982

D／パオス, ケンウッド・コーポレートデザイン部

356
スポーツ用品会社
ヨネックス
東京
1986

D／上條喬久

357
デパート
伊勢丹
東京
1985

CD／土屋耕一
AD／戸田正寿
D／鈴木 守
DF／T・Room, 戸田事務所
店名ロゴタイプのイニシャル"I"を大きくシンボル化し, 紺と黄がカラーテーマに使われている。白地タイプ, 黄色地タイプ, 食品売場用の茶色ベースの3種がある。

354

355

354
Fashion Boutique
Shiseido Boutique The Ginza
Tokyo, Japan
1985

D／Masayoshi Nakajo

355
Audio Appliances Manufacturer
Kenwood Corporation
Tokyo, Japan
1982

AD／Motoo Nakanishi
D／Paos, Inc., Corporate Design Dept., Kenwood Corporation

356
Sporting Goods Company
Yonex Co., Ltd.
Tokyo, Japan
1986

D／Takashi Kamijyo

357
Department Store
Isetan Department Store Co., Ltd.
Tokyo, Japan
1985

CD／Koichi Tsuchiya
AD／Masatoshi Toda
D／Mamoru Suzuki
DF／T Room, Toda Studio
The initial "I" of the logotype for the shop name is made into a large symbol, using the colors of dark blue and yellow. There are the three types of white background, the yellow background and the brown base color food counters.

356

357

358

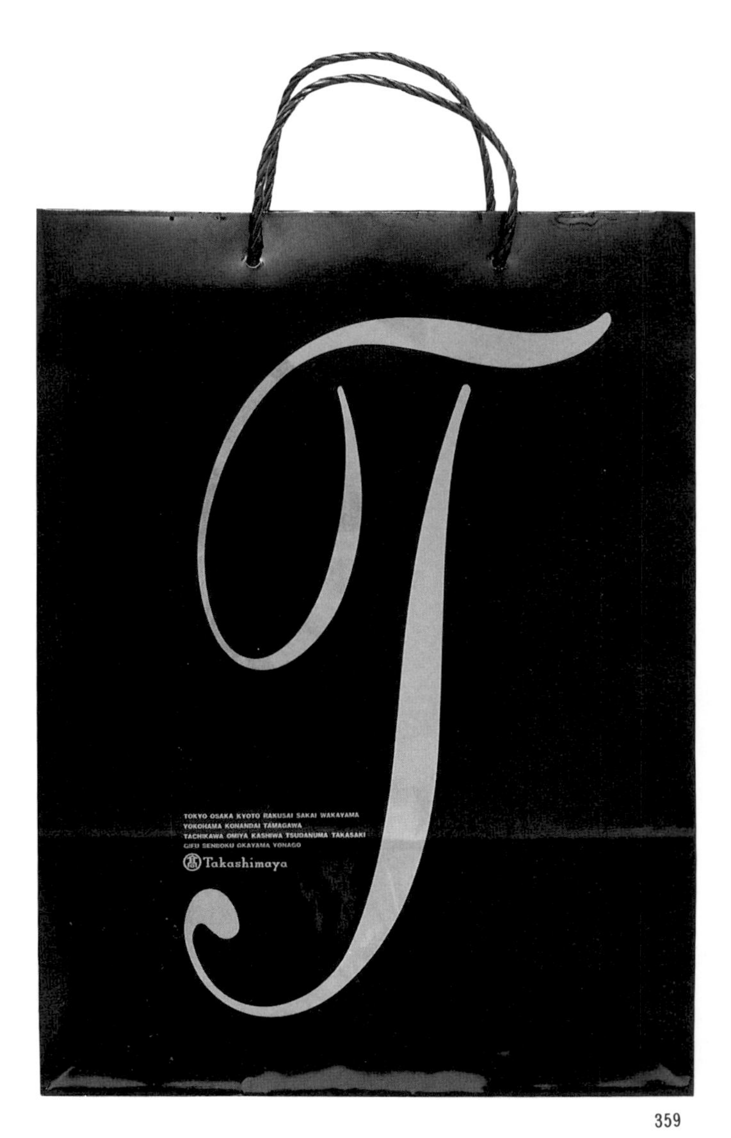

359

360

358
美術館
目黒区美術館
東京
1987

D／矢萩喜従郎
DF／キジュウロウ・ヤハギ

359
デパート
高島屋
東京

D／駒形敬三
DF／宣研

360
洋菓子会社
リュバンドール
神戸市
1981
D／ヘルムート・シュミット
Planner／西尾捷太
A／JMS

358
Art Museum
Meguro Museum of Art, Tokyo
Tokyo, Japan
1987

D／Kijuro Yahagi
DF／Kijuro Yahagi Co., Ltd.

359
Department Store
Takashimaya Co., Ltd.
Tokyo, Japan

D／Keizo Komagata
DF／Senken Co., Ltd.

360
Confectionary
Ruban d'or
Kobe City, Japan
1981

D／Helmut Schmid
Planner／Shota Nishio
A／Japan Marketing Service Co., Ltd.

361

362

361
ブティック
青山商品研究所(Again)
東京
1987

AD/富樫大輔
D/横山忠正
DF/横山忠正事務所

362
CM制作会社
CMらんど
東京
1982

AD/浅葉克己
D/大木理人
I/長 新太

361
Boutique
Aoyama Research Institute of
Fashion Merchandising Co., Ltd.
(Again)
Tokyo, Japan
1987

AD/Daisuke Togashi
D/Tadamasa Yokoyama
DF/Yokoyama Office

362
CM Production
CM Land Co., Ltd.
Tokyo, Japan
1982

AD/Katsumi Asaba
D/Rijin Oki
I/Shinta Cho

363

363
アパレル・ブランド
キャスネル(ヴィヴィアン・ウェストウッド)
イギリス, ロンドン
1987

D／ヴィヴィアン・ウェストウッド

364
玩具・雑貨店
ビニール
キデイランド
東京
1987

DF／ライト・パブリシティ

365
"グレース・ブラザーズ"
グレース・ブラザーズ・パーティ・リミテッド
オーストラリア, シドニー
1987

D／ケン・ケイトー
DF／ケイトー・デザイン・インコーポレイテッド

363
Apparel Brand
Casnell Ltd. (Vivienne Westwood)
London, England
1987

D／Vivienne Westwood

364
Toy Boutique
Plastic
Kiddy Land Co., Ltd.
Tokyo, Japan
1987

DF／Light Publicity Co., Ltd.

365
"Grace Bros."
Grace Bros.Pty. Ltd.
Sydney, Australia
1987

D／Ken Cato
DF／Cato Design Inc.

364

365

366

367

366
レコード,テープ用バッグ
ビニール
ガスライト・レコード・アンド・テープ
オーストラリア,メルボルン
1983

D/フィリップ・エレット
DF/コゾリーノ/エレット

367
ブティック
ビニール
ビームス
東京
1984

AD/設楽 洋
D/鳥塚 寧
DF/ビームス・クリエイティブ,新光

366
Records and Tapes Bag
Plastic
Gaslight Records & Tapes
Melbourne, Australia
1983

D/Philip Ellett
DF/Cozzolino / Ellett

367
Boutique
Plastic
Beams Co., Ltd.
Tokyo, Japan
1984

AD/Yo Shitara
D/Yasushi Toritsuka
DF/Beams Creative Inc., Shinko Co., Ltd.

368

369

368
バッグ・ショップ
紙，ポリエチレン
ザ・ギンザ（ラ・バガジェリー）
東京
1987

D／太田和彦

369
スポーツ用具店
ポリエチレン
セントクリストファー
東京
1987

D／野口正治
DF／スタジオ・コムテック

368
Bag Shop
Paper, Plastic
Shiseido Boutique The Ginza
(La Bagagerie)
Tokyo, Japan
1987

D／Kazuhiko Ota

369
Sporting Goods Shop
Plastic
St. Chrictopher Inc.
Tokyo, Japan
1987

D／Masaharu Noguchi
DF／Studio Comtech

370
子供服店
ビニール
リトルアンデルセン(チャビー・ギャング)
東京
1986

D/村上義行

371
ぬいぐるみ店
ポリエチレン
ハニー(カドリー・ブラウン)
東京
1984

AD/佐久間由紀子
D/山崎真由美
DF/インタレスト
カドリー・ブラウンは，キャラクター“テディ・ベア”の専門店。かわいらしさだけでなく，テディ・ベアのオーソドックスでトラディッショナルなイメージでデザインされた。

372
書店
ビニール
鹿六(メディア・ショップ)
京都
1987

D/田中岑也

370
Children's Clothing Shop
Plastic
Little Andersen Co., Ltd. (Chaby Gang)
Tokyo, Japan
1986

D/Yoshiyuki Murakami

371
Stuffed Toy Shop
Plastic
Honey Co., Ltd. (Cuddly Brown)
Tokyo, Japan
1984

AD/Yukiko Sakuma
D/Mayumi Yamazaki Interest Inc.
"Cuddly Brown" is a shop selling Teddy Bears. Designed not only to show the cute image but also the orthodox, traditional image of Teddy Bears.

372
Bookstore
Plastic
Karoku Co., Ltd. (Media Shop)
Kyoto, Japan
1987

D/Mineya Tanaka

370

371
372

373

373
手織物ショップ
紙, ビニール
バーバラ・ペテルカ
ユーゴスラヴィア, リュブリアナ
1988

D／エディ・ベルク
P／ドゥラガン・アリグラー
DF／スタジオ KROG
手織商品を扱っている店のC. I.システムの一環としてつくられたショッピングバッグ。同店の企業カラーが使われている。

374・375
"ジャイアント・ストアーズ"
ジャイアント・ストアーズ・リミテッド
ニュジーランド, オークランド
1987

D／ケン・ケイトー
DF／ケイトー・デザイン・インコーポレイテッド

376
カーテン・家具会社
ビニール
ダイバ・ソフト・ファーニシングズ
カナダ, モントリオール
1984

D／ロルフ・ハーダー
DF／ロルフ・ハーダー・アンド・アソシエイツ

373
Weaving Supplier
Paper, Plastic
Barbara Peterca, Hand Weaving
Ljubljana, Yugoslavia
1988

D／Edi Berk
P／Dragan Arrigler
DF／Studio KROG
Shopping bag are a part of the corporate identity of this small firm, a weaving supplier. Colors are consistent with the company image.

374・375
"Giant Stores"
Giant Stores Limited
Auckland, New Zealand
1987

D／Ken Cato
DF／Cato Design Inc.

376
Soft Furnishings Company
Plastic
Diva Soft Furnishings
Montréal, Canada
1984

D／Rolf Harder
DF／Rolf Harder & Assoc., Inc.

GIANT

374

375

376

377

378

377
宝石店
宝石の富士屋
大阪
1983

D／高田雄吉
DF／アイ・エフ・プランニング
CIの一環として，シンボルマーク，ロゴタイプ，店名だけを使ってデザインされたショッピングバッグ。銀のホットスタンプは光を反射するときと吸収するときがあり，マークは背景に対してネガになったり，ポジになったりする。“Jewelry Fujiya”のイニシャルをシンボル化したJとFのマークは，互いにネガとポジの関係にデザインされた。

378
ホテル
札幌パークホテル
札幌市
1987

D／土屋知行
DF／イメージ・イン・スタジオ

377
Jeweler's Shop
Jewelry Fujiya
Osaka, Japan
1983

D／Yukichi Takada
DF／I.F. Planning Co., Ltd.
A shopping bag and the wrapping paper designed using only a symbol mark, logo and shop name as one item for the C.I. project promotion. The silver stamp reflects or absorbs light, and the mark will have a negative image or positive image accordingly to the background. The two characters “J” and “F”, which form the mark of Jewelry Fujiya, are designed to alternate between the negative and positive images.

378
Hotel
Sapporo Park Hotel
Sapporo City, Japan
1987

D／Tomoyuki Tsuchiya
DF／Image Inn Studio Co., Ltd.

379

380

381

379
レコード店
ビニール
バナナレコード
名古屋
1986

D／鈴木 勝
I／鈴木かずえ

380
ファッション・ブティック
ビニール
ブティック・ギルズ
スイス
1987

D／カール・ドメニック・ガイスビューラー

381
ブティック
ビニール
ピンクバス
東京
1986

D／松本伊代

379
Record Shop
Plastic
Banana Record
Nagoya City, Japan
1986

D／Masaru Suzuki
I／kazue Suzuki

380
Fashion Boutique
Plastic
Boutique Gilles
Switzerland
1987

D／Karl Domenic Geisbühler

381
Boutique
Plastic
Pink Bus
Tokyo, Japan
1986

D／Iyo Matsumoto

382
"グリーン・キング"
ビニール
グリーン・キング・アンド・サン
イギリス，サファク州ベリー・セント・エドモンズ
1982

D／デーヴィッド・ポックネル
I／ロン・マーサー
DF／ポックネル・アンド・カンパニー
グリーン・キングのロゴはショッピングバッグを含む数々のディスプレイ・アイテム，すなわち酒場のサインボードやコースター，バンの外装などにも使われている。

383
デパート
ビニール
ハロッズ・ナイトブリッジ・リミテッド
イギリス，ロンドン

D／ミナーレ・タタースフィールド・プラス・パートナーズ

384
陶磁器・ガラス食器店
ビニール
ローゼンタール
東京
1978

AD／リタ・ギュンターマン
D／ローゼンタール

382
"Greene King"
Plastic
Greene King & Son PIC.
Bury St. Edmunds, Suffolk, England
1982

D／David Pocknell
I／Ron Mercer
DF／Pocknell & Co.
Colored line Illustration of Greene King logo—which appeared on display items such as public house signs ; Beer Mats ; Van Livery and Shopping Bags.

383
Department Store
Plastic
Harrods Knightbridge Ltd.
London, England

D／Minale Tattersfiend+Partners

384
Tableware Shop
Plastic
Rosenthal
Tokyo, Japan
1978

AD／Rita Güntermann
D／Rosenthal

382

383

384

385

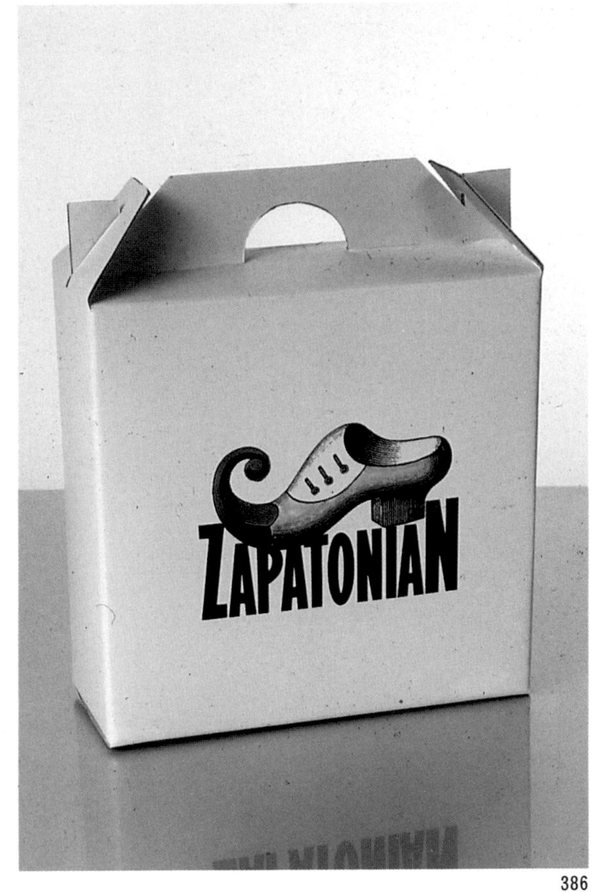

386

387

388

385
アパレル・ブティック
ポリエチレン
ワールド(ジ・エンポリアム)
大阪
1987

AD/藤本真樹
D/岡田 哲
DF/Heaven

386
靴店
ザパトニアン・シューズ
スペイン・マドリード
1978

D/フェルナンド・メディーナ
DF/メディーナ・デザイン

387
ブティック
ビニール
フリーウェイ(フリーウェイ428)
横浜
1983

D/小島升夫
DF/フリーウェイ・デザイン・ワーカーズ

388
バラエティ・ショップ
ポリプロピレン
ウィークエンズ
東京
1987

D/近藤絹衣
DF/ウィークエンズ
A/エリス

385
Apparel Boutique
Plastic
World Co., Ltd. (The Emporium)
Osaka, Japan
1987

AD/Maki Fujimoto
D/Tetsu Okada
DF/Heaven

386
Shoe Store
Zapatonian Shoes
Madrid, Spain
1978

D/Fernando Medina
DF/Medina Design

387
Boutique
Plastic
Freeway Co., Ltd. (Freeway 428)
Yokohama, Japan
1983

D/Masuo Kojima
DF/Freeway Design Workers

388
Variety Shop
Plastic
Weekends
Tokyo, Japan
1987

D/Kinue Kondo
DF/Weekends
A/Elice Co., Ltd.

389

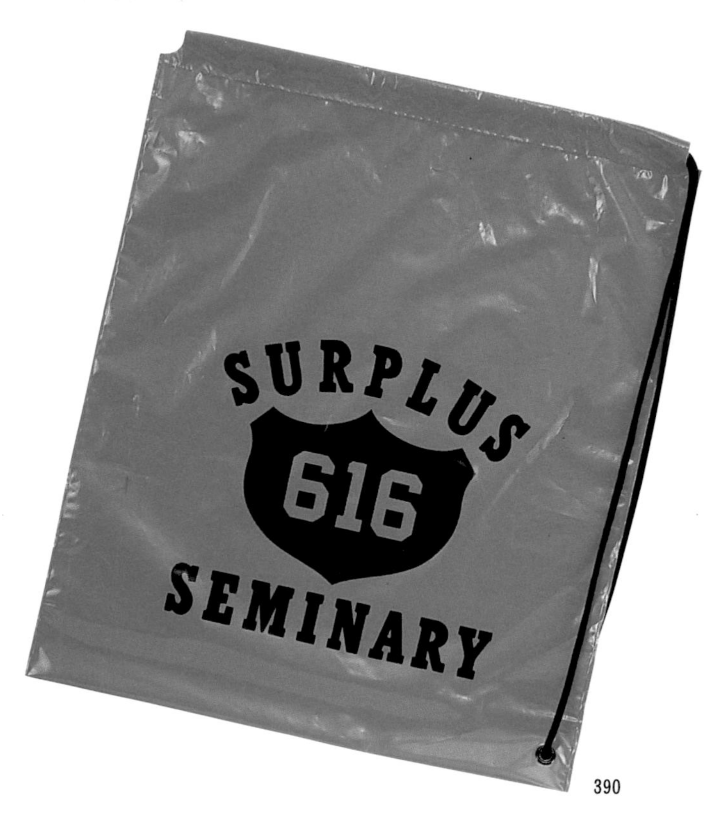

390

391

392

389
洋装・雑貨店
ビニール
グラマー
大阪

D／グラマー

390
ブティック
ビニール
サープラス
東京
1987

D／老川輝久
DF／新東

391
ブティック
ビニール
ガスト・ハウス
東京
1987

D／菊池光治

392
アパレル・ブティック
ビニール
チャンプ・カンパニー
神戸市
1986

D／キモリ産業

389
Clothing and Missellaneous
Goods Store
Plastic
Glamour
Osaka, Japan

D／Glamour

390
Boutique
Plastic
Surplus Inc.
Tokyo, Japan
1987

D／Teruhisa Oikawa
DF／Shinto Co., Ltd.

391
Boutique
Plastic
Gast Haus
Tokyo, Japan
1987

D／Koji Kikuchi

392
Apparel Boutique
Plastic
Champ Company
Kobe City, Japan
1986

D／Kimori Sangyo Co., Ltd.

393
トイ・デパート
博品館トイ・パーク
東京
1982

D／博品館トイ・パーク

393
Toy Store
Hakuhinkan Toy Park
Tokyo, Japan
1982

D／Hakuhinkan Toy Park

393

394

394
テイクアウトのデリカテッセン
レストラン・アソシエイツ（セイバリーズ）
アメリカ，ニューヨーク
1985

D／トム・ガイスマー
DF／シャマイエフ・アンド・ガイスマー・アソシエイツ
テイクアウトのデリカテッセン"セイバリーズ"のロゴ・デザインは"アメリカの起源"がテーマとなっている。文字とイラストはアメリカの刺繍に見られるクロス・ステッチを応用したもの。

394
Take-out Restaurant/
Delicatessen "Savories"
Restaurant Associates
New York, USA
1985

D／Tom Geismar
DF／Chermayeff & Geismar Associates
The logo for Savories, a take-out restaurant/delicatessen, was designed to reflect an American origin. The type and illustration are based on the cross stitch look of an American sampler.

395
婦人服店
アイドル
東京
1987

DF／ザ・デザイン・アソシエイツ
A／丹青社
ロゴ"O"と店のある街のシンボル"天使像"をエレメントとしてデザイン。「南ヨーロッパの伝統＆ノスタルジーとして展開する」というショップ・コンセプトをもつ店の個性の主張が制作のねらい。

395
Ladies' Clothing Shop
Idol Co., Ltd.
Tokyo, Japan
1987

DF／The Design Associates Co., Ltd.
A／Tanseisha Co., Ltd.
Designed using the log "O" and the town's "Angel" symbol as the basic elements, the aim is to emphasize the uniqueness of the shop's concept-"to develop a feeling of nostalgia for the traditions of Southern Europe and products in the form of an urban casual fashion".

395

396

396
女王誕生日祝賀
ザ・クイーンズ・バースデー・コミッティ
イギリス, ロンドン
1986

D／マーヴィン・カーランスキー
I／ウォルク・スポール
DF／ペンタグラム・デザイン
1986年4月21日，クィーン・エリザベス2世の60歳の誕生日を祝うためにデザインされたシンボルが，ポスター，T-シャツ，マグカップ，ショッピングバッグなど，いろんなグラフィック・マテリアルに使われた。

396
Queen's Birthday Celebration
The Queen's Birthday Committee
London, England
1986

D／Mervyn Kurlansky
I／Wolf Spoerl
DF／Pentagram Design
A symbol designed to celebrate the 60th Birthday of Queen Elizabeth II on 21 April 1986. The symbol was also applied to posters, T-shirts, mugs shopping bags, and other graphic material.

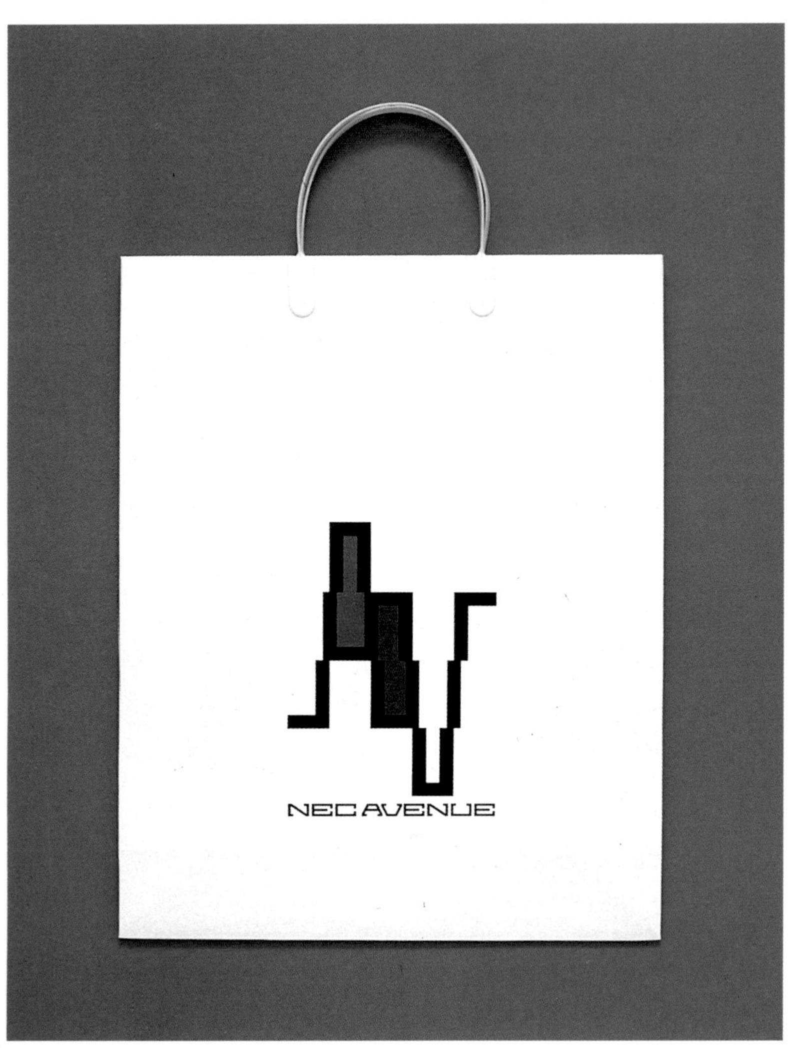

397

397
AV 会社
NEC アベニュー
東京
1987

D／中村政久, 鈴木善博
DF／ビーバイスタジオ
A／電通

398
アパレル・ブティック
ラパス
神戸市

D／井上雄史

397
Audio Visual Components Company
NEC Avenue Co., Ltd.
Tokyo, Japan
1987

D／Masahisa Nakamura, Zempaku Suzuki
DF／B・Bi Studio Inc.
A／Dentsu, Inc.

398
Apparel Boutique
Lapaz
Kobe City, Japan

D／Yuji Inoue

398

399
化粧品会社'88春のキャンペーン
資生堂
東京
1988

AD／田島一夫
D／伊藤 満
春期キャンペーン用のハンディバッグ。口紅を写真撮影し，粗めの万線スクリーンでイラスト風に表現し，"LIVE LIP"のロゴと組み合わせた。フィルムに裏刷りしたものを紙にラミネートすることによって，つやを出している。

399
Cosmetics Company '88 Spring Campaign
Shiseido Co., Ltd.
Tokyo, Japan
1988

AD／Kazuo Tajima
D／Mitsuru Ito
A handy bag for the "Spring Campaign". A photo of lip stick, modified to look like an illustration by means of a coarse screen, makes up the "Live Lip" logo. Printed on film and then laminate it on paper, it gives glossy surface.

399

400

400
ブティック
B×3·Node館
東京
1986

AD／丸山宏子
D／木内 哲
I／唐木 潤
DF／TB-FMA

400
Boutique
Node House, B-By-Three Co., Ltd.
Tokyo, Japan
1986

AD／Hiroko Maruyama
D／Satoshi Kiuchi
I／Jun Karaki
DF／TB-Fashion Marketing Agency Inc.

401

402

401
家電メーカー
三洋電機
大阪
1986

AD／ハワード・ヨーク, ユージン・グロスマン
D／ハワード・ヨーク
DF／アンスパック・グロスマン・ポーチュガル・インコーポレイテッド
サンヨーの新しいC. I.システムにのっとってデザインされたもの。このデザインは世界中の電気製品販売店に行き渡っている。

402
バラエティ・ショップ
ビニール
ナイス・クリエーション(デタント)
ニューヨーク, 東京
1987

AD／渡辺栄次, 河本義成
D／河島竜作
I／寺井 潔
DF／ナイス・クリエーション

401
Electric Appliances Manufacturer
Sanyo Electric Co., Ltd.
Osaka, Japan
1986

AD／Howard York, Eugene Grossman
D／Howard York
DF／Anspach Grossman Portugal Inc.
Shopping Bags and wrapping paper have been designed to reflect Sanyo's new identification system, which is used worldwide in retail electric product stores.

402
Variety Shop
Plastic
Nice Creation Co., Ltd. (Detente)
New York, USA ; Tokyo, Japan
1987

AD／Eiji Watanabe, Yoshinari Kawamoto
D／Ryusaku Kawashima
I／Kiyoshi Terai
DF／Nice Creation Co., Ltd.

403
デパート
西武百貨店(Seibu Sports)
東京
1979

D/浅葉克己

404
デパート
藤丸
帯広市
1979

D/草刈 順

403
Department Store
The Seibu Department Stores Co., Ltd. (Seibu Sports)
Tokyo, Japan
1979

D/Katsumi Asaba

404
Department Store
Fujimaru Co., Ltd.
Obihiro City, Japan
1979

D/Jun Kusakari

403

404

405

405
レコード会社
フォーライフレコード
東京
1986

D／上條喬久

405
Record Company
For Life Record Co., Ltd.
Tokyo, Japan
1986

D／Takahisa Kamijyo

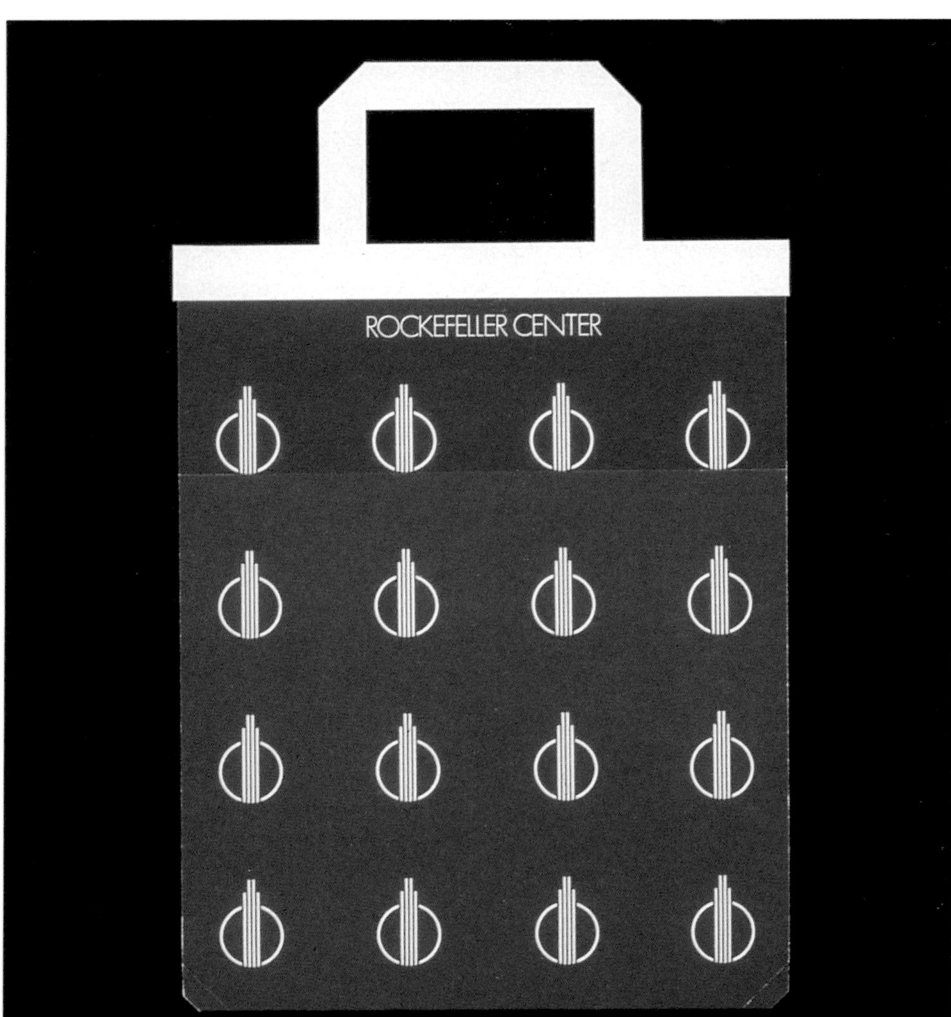

406

406
ロックフェラー・センター
アメリカ, ニューヨーク

D／アイヴァン・シャマイエフ, トム・ガイスマー, ステッフ・ガイスビューラー
DF／シャマイエフ・アンド・ガイスマー・アソシエイツ

406
Rockefeller Center
New York, USA

D／Ivan Chermayeff, Tom Geismar, Steff Geissbuhler
DF／Chermayeff & Geismar Associates

407
ソックス専門店
ビニール
サムシング・アフット
アメリカ, テキサス州ダラス
1986

D／ジャック・サマーフォード
DF／サマーフォード・デザイン
ダラスの"サムシング・アフット"は, ソックスの専門店。ソックスを模したアポストロフィーのついたロゴはショッピングバッグをはじめ, ステーショナリー, サインボードなどにも使われている。

408
ファッション・ブティックのクリスマス・キャンペーン
ザ・ギンザ
東京
1986

D／川崎修司
DF／ヘッズ
クリスマスにふさわしいように, レギュラーのバッグとは変えて, 山型のマーク, チェックの包装紙(通常はグレー, ブラウン1色)を赤と緑で切り変え, 少し華やかに見せようという意図。

407
Socks and Hosiery Shop
Plastic
Something's Afoot
Dallas, Tex., USA
1986

D／Jack Summerford
DF／Summerford Design, Inc.
Something's Afoot is a retail boutique which specializes in socks and hosiery for both men and women. The logo capitalizes on the recognizable sock shape and the need for an apostrophe in the company name. Applications vary from stationery to shopping bags and signage.

408
Fashion Boutique Christmas Campaign
Shiseido Boutique The Ginza
Tokyo, Japan
1986

D／Shuji Kawasaki
DF／Heads Inc.
For preparing products in a more gay design to match the mood of the coming Christmas season, a modification is made to bags and wrapping paper already in stock through the introduction of the mountain-shaped motif. A change of the colors of the check pattern on the wrapping paper, to red and green, was also added.

407

408

409

410

411

409
レジャークラブ
オレンジ・ビレッジ・クラブ
神奈川県
1983

D／草刈 順

410
デパート
阪神百貨店
大阪
1978

D／草刈 順
A／アルテリア

411
量販店
ミドリヤ
東京
1978

D／草刈 順

409
Leisure Club
Orange Village Club Co., Ltd.
Kanagawa Pref., Japan
1983

D／Jun Kusakari

410
Department Store
Hanshin Department Store
Co., Ltd.
Osaka, Japan
1978

D／Jun Kusakari
A／Arteria Co., Ltd.

411
Superstore
Midoriya Co., Ltd.
Tokyo, Japan
1978

D／Jun Kusakari

412

413

412
ファッション・ブランド"パーキー・ジーン"
ザ・ギンザ
東京
1986

D／仲條正義

413
ケーキ店
パウンド・ハウス
大阪

D／パウンド・ハウス

412
Fashion Brand "Perky Jean"
Shiseido Boutique The Ginza
Tokyo, Japan
1986

D／Masayoshi Nakajo

413
Confectionery
Pound House
Osaka, Japan

D／Pound House

414

414•415
個展のためのショッピングバッグ
東京デザイナーズスペース
東京
1987

D／川崎修司, 高橋新三
P／能津喜代房
DF／ヘッズ
東京デザイナーズスペースでの個展のために，取引のある10社を選んで自由に制作したもの。

414•415
Shopping Bag Works for the One-man Exhibit
Tokyo Designer's Space
Tokyo, Japan
1987

D／Shuji Kawasaki, Shinzo Takahashi
P／Kiyofusa Nozu
DF／Heads Inc.
Optionally designed works on selected items for ten clients for one-man exhibiting at the Tokyo Designer's Space.

415

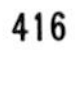

416

417

416
ブティック
ビニール
イング・セブン(ポッシュ・ボーイ)
東京
1987

D/矢沢義夫
I/薙野たかひろ

417
バラエティ・ショップ
パピルス・カンパニー
東京
1987

D/平井美樹
I/飯田 淳
DF/パピルス・カンパニー

416
Boutique
Plastic
Ing Seven Co., Ltd. (Posh Boy)
Tokyo, Japan
1987

D/Yoshio Yazawa
I/Takahiro Nagino

417
Variety Shop
Papyrus Co., Ltd.
Tokyo, Japan
1987

D/Miki Hirai
I/Jun Iida
DF/Papyrus Co., Ltd.

INO.
DECONTRACTE ET ESSENTIEL
INO
DECONTRACTE ET ESSENTIEL

418

ALL STAR FAIR
ELEX
SANYO
ALL STAR FAIR
ELEX
SANYO

419

418
ファッション・ブランド"イノー"
ジュン
東京
1986

AD／ジュン

418
Fashion Brand Shopping Bag
Jun Co., Ltd.
Tokyo, Japan
1986

AD／Jun, Co., Ltd.

419
家電メーカーの展示会用バッグ
三洋電機
大阪
1988

AD／須波 光
DF／プロジェクト・コム
A／電通
1988年のプロ野球オールスターゲームの冠スポンサーに決まった三洋電機の商品展示会用のショッピングバッグ。白と赤を基本色にして，展示会名称の"オールスター・フェア"のロゴをあしらっている。

419
Electric Appliances
Manufacturen Exhibition Bag
Sanyo Electric Co., Ltd.
Osaka, Japan
1988

AD／Hikaru Sunami
DF／Project COM
A／Dentsu, Inc.
This shopping bag was designed for a fair held by the Sanyo Electric Company, a sponsor for the 1988 Japan Professional Baseball All-star Game. Using red and white as its basic colors, the bag emblazons the name of the fair in a bold and sporting design.

420

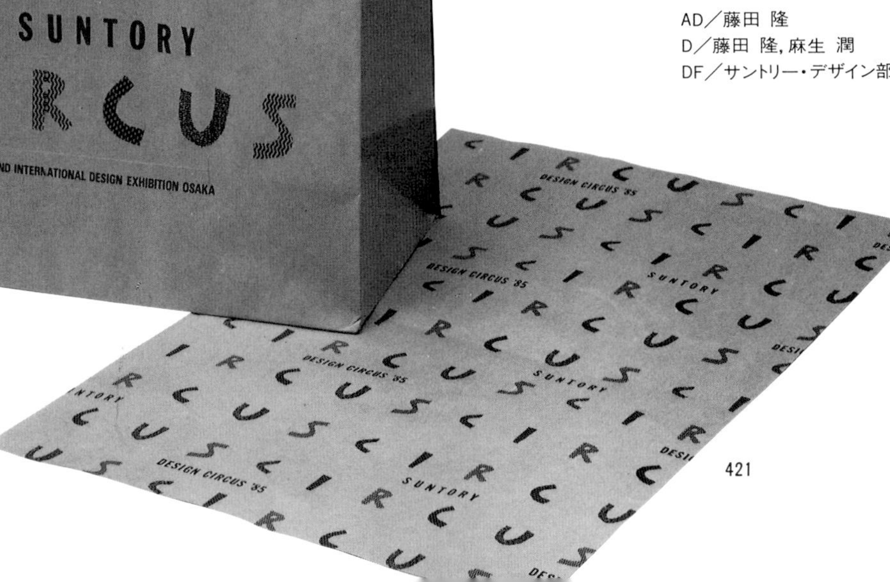

421

420
特製品ストアーの衣料袋
ポリエチレン
アップズ・アンド・ダウンズ
アメリカコネチカット州エンフィールド
1987

AD／ロバート・ガーシン
D／ラルフ・エーインガー
DF／ロバート・ガーシン・アソシエイツ

420
Garment Bag of a Specialty Retail Store
Plastic
Ups and Downs
Enfield, CT, USA
1987

AD／Robert Gersin
D／Ralph Ehinger
DF／Robert P. Gersin Associates, Inc.

421
大阪デザイン博の洋酒会社パビリオンのバッグ
サントリー
大阪
1985

AD／藤田 隆
D／藤田 隆，麻生 潤
DF／サントリー・デザイン部

421
Bag for Liquor Company Pavillion, Osaka International Design Exhibition 1985
Suntory Limited
Osaka, Japan
1985

AD／Takashi Fujita
D／Takashi Fujita, Jun Aso
DF／Suntory Limited

BLUE COLUMBINES
AND
GOLDEN POPPIES
ACROSS
THE
L·A·N·D
NATIONAL
COUNCIL
OF STATE
GARDEN
CLUBS
INC.
53RD
MAY
23-26
1982
ANNUAL CONVENTION
CENTURY PLAZA
LOS ANGELES

422

Furest

423

Furest

423

422
ガーデン・クラブ
ナショナル・カウンシル・オブ・ステート・ガーデン・クラブ，Inc.
アメリカ，ロサンゼルス
1982

D／スタン・エヴァンソン
DF／スタン・エヴァンソン・デザイン

423
メンズウェア・ショップ
フレスト
スペイン，バルセロナ
1978

D／アメリカ・サンチェス
"フレスト"は，バルセロナの商業地域に4店舗をもつメンズファッションの店。バッグの表と裏には，違うイラストが描かれているが，どちらもトロピカルなイメージになっている。

422
"Blue Columbines and Golden Across the Land"
National Council State Garden Clubs, Inc.
Los Angeles, USA
1982

D／Stan Evenson
DF／Stan Evenson Design

423
Men's Wear Shop
Furest
Barcelona, Spain
1978

D／America Sanchez
"Furest" is a men's clothing shop with four branches in the Barcelona commercial area. This bag has two different illustrations on its front and back, both inspired by tropical landscapes.

424

424
スーパーマーケット
イルマ・スーパーマーケット
デンマーク，コペンハーゲン
1984

D／フレミング・ニールセン
DF／ニールセン・アンド・ベイリー
このショッピングバッグはデンマーク・デザイナーズ・オーガナイゼイション，IDDのメンバーによってつくられたシリーズの中の1点。デザイナーに与えられた指示は，黒で印刷するということだけであり，それ以外はすべて任された。WWF（世界野生動物保護基金）の運動への賛助活動として制作されたもの。

424
Supermarket
Irma Supermarket
Copenhagen, Denmark
1984

D／Flemming Nielsen
DF／Nielsen & Baillie
The client's only condition for this shopping bag design was that the final print be black. Designed for the WWF (World Wildlife Foundation).

425
温泉郷
岐阜県吉城郡上宝村役場
岐阜県
1986

AD／石田 隆
D／石川 和市
I／石田 隆
奥飛騨温泉郷の露天風呂や伝統行事に，周辺の山岳観光をアレンジし，黒一色で"素朴さ"を表現している。

426
雑貨店
ビニール
シン＆カンパニー
東京
1986

D／稲垣 晋
DF／シン＆カンパニー

425

426

427 428

427
古美術店
古美術・田澤
京都
1980

I／木田安彦

428
レストランのワイン袋
南欧料理・グルモン
京都
1980

I／木田安彦
レストランのワイン袋で，南欧風のモチーフを，絵地図的に描いた。

425
Hotspring Resort
Kamitakara-mura Village Office
Gifu Pref., Japan
1986

AD／Takashi Ishida
D／Kazushi Ishikawa
I／Takashi Ishida
An outdoor open-air bath and local traditional events in the Okuhida hotspring resort area and mountain sighteeing spots surrounding the area are arranged in shade of black only to present an unsophisticated atmosphere.

426
Grocery Store
Plastic
Shin & Company Ltd.
Tokyo, Japan
1986

D／Shin Inagaki
DF／Shin & Company Ltd.

427
Antique Shop
Tazawa
Kyoto, Japan
1980

I／Yasuhiko Kida

428
Restaurant Wine Bottle Bag
Gourmand
Kyoto, Japan
1980

I／Yasuhiko Kida
A bag for wine with a southern European motif, drawn in an illustrated map style.

429
Umbrella Shop
Great Britain Corporation
(Gene Kelly)
Tokyo, Japan
1986

D／Kunio Tokunaga
A／Fuji Paper Industry Co., Ltd.
This bag has been designed with pictures from newspapers on dark yellow peper, lending a nostalgic air and creating an image of the late 1920's.

430
Apparel Boutique
D'Letes Club (Little Sleek Bean)
Tokyo, Japan
1984

D／Makiko Hayami
DF／Mugen Corporation

429

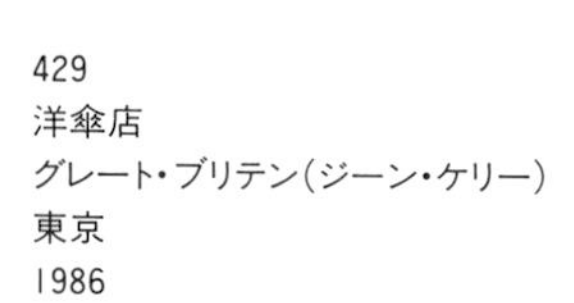
429
洋傘店
グレート・ブリテン(ジーン・ケリー)
東京
1986

D／徳永邦夫
A／フジ紙工
昭和初期(1920年代後半)のイメージで，やや黄ばみ気味の紙に，当時を想像させるレトロ感覚の新聞紙を絵柄にしている。

430
アパレル・ブティック
デライツ・クラブ(リトル・スリーク・ビーン)
東京
1984

D／速見牧子
DF／夢幻

430

431
バラエティ・ショップの'87クリスマス・キャンペーン
ソニー・プラザ
東京
1987

D／渡辺通正
最もポピュラーなクリスマス・カラーであるグリーンとレッドを使用。もみの木と雪をモチーフに，"アットホームな心暖まるクリスマス"のイメージを表現するため，フリーハンドのやわらかな線描きのイラストレーションをあしらっている。

431
Variety Shop '87 Christmas Campaign
Sony Plaza Co., Ltd.
Tokyo, Japan
1987

D／Michimasa Watanabe
Designed using the popular Christmas colors of red and green. The illustration work has been done in soft, free hand lines, in order to create a warm Christmas image with the motives of fir trees and snow.

431

432

432
デパートのクリスマスセール
松坂屋（上野店）
東京
1981

D／勝俣和良
DF／松坂屋（上野店）
クリスマスムードを盛り上げ，楽しい雰囲気の中で買い物ができるような装飾，演出効果の一環として制作された。ローカルのクリスマス用バッグとして社内デザイナーによるコンペで募ったもの。

432
Department Store Christmas Sale
Matsuzakaya Co., Ltd. (Ueno Store)
Tokyo, Japan
1981

D／Kazuyoshi Katsumata
DF／Matsuzakaya Co., Ltd. (Ueno Store)
A bag to play a key role, together with the ornaments and dramatic presentations in creating the Christmas mood. This work won a prize for a "Christmas carrier bag for local stores" at a competition held for in-house designers of the company.

433
量販店のウィンター・セール
田原屋
川崎市
1987

D／東京グラフィック

434
メンズ・ブティック
メンズショップ・トラヤ
大阪
1986

D／メンズショップ・トラヤ

435
カジュアル・ショップ
ロッキー
大阪
1980

I／横尾忠則
DF／パノラマ・アドバタイジング

433
Superstore Winter Sale
Tawaraya Co., Ltd.
Kawasaki City, Japan
1987

D／Tokyo Graphic Co., Ltd.

434
Men's Boutique
Men's Shop Toraya
Osaka, Japan
1986

D／Men's Shop Toraya

435
Casual Clothing Shop
Rocky Co., Ltd.
Osaka, Japan
1980

I／Tadanori Yokoo
DF／Panorama Advertising Inc.

433

434

435

A STORE ROBOT
LONDON NEWYORK TOKYO NAGOYA OSAKA GIFU
436

437

PIN-UP SUMMER
PIN-UP SUMMER
PIN-UP·SUMMER
438

436
ブティック
ビニール
ア・ストア・ロボット
名古屋市
1987

D／伊東佐英子
DF／ア・ストア・ロボット
ロック少年たちを意識して制作された。客層は，10代後半～20代前半のロック少年たちが多く，彼らに人気のあるドクロ模様のＴシャツやアクセサリーに着目して，ドクロのバッグにしたもの。

437
ブティック
ビニール
ナイス
東京
1986

D／木村力三
アメリカ1950年代へのあこがれの気持をこめて，イラストをバッグ全体にちりばめ，さらに手で持つより肩にかけてリュックサックになるショルダー型ということで，遊び感覚を大切にデザインされた。

438
市販用"ピンナップ・サマー"
ポリエチレン
リリック
東京
1987

D／稲垣慶子
DF／プラネット
「夏の想い出を心の中にピンナップする」というコンセプトで白い砂浜，コバルトブルーの海，白壁にうつる影など，想い出と現実の交差するノスタルジックな風景を，シルエットの男女の世界で表現。

439
染毛剤会社
ホーユー
名古屋
1984

D／山内瞬葉
I／山内瞬葉
P／勝田安彦
「ハート型ウォッチプレゼントセール」用につくられたバッグ。イラストをピントをずらして複写してカラー分解し，正確に複写したイラストの線画部分を合成したもの。

440
婦人用皮製品ブティック
ラリオ
ホンコン
1983

AD／アラン・チャン
D／アラン・チャン，アルヴィン・チャン
DF／アラン・チャン・デザイン
"ラリオ"は，ヨーロッパ直輸入の靴やハンドバッグ類を販売している婦人用皮革製品のブティック。透明なチューブの把手はシンデレラのガラスの靴がコンセプトのベースになっている。

436
Boutique
Plastic
A Store Robot Co., Ltd.
Nagoya City, Japan
1987

D／Saeko Ito
DF／A Store Robot Co., Ltd.
This shop was established with young rockers in ming. Most customers are usually in the age range of late teens to early twenties ; the most popular products with the young rockers are T-shirts and accessories emblazoned with a skull design.

437
Boutique
Plastic
Nice & Company
Tokyo, Japan
1986

D／Rikizo Kimura
An illustration inlayed over the whole surface of the bag, giving an American 50's look. As it is more of a shoulder bag than a hand bag, it can also function as a ruck sack. It has been designed more for casual use.

438
Shopping Bag "Pin-up Summer"
Plastic
Lyric Co., Ltd.
Tokyo, Japan
1987

D／Keiko Inagaki
DF／Planet, Inc.
Designed with "Pin-up Summer" as its concept, conjuring up with the silhouette of man and woman, nostalgic memories of a world of white sandy bays, cobalt blue seas, shadows on a white wall...

439
Hairdye Company
Hoyu Co., Ltd.
Nagoya City, Japan
1984

D／Shunyo Yamauchi
P／Yasuhiko Katsuda
A bag designed for the "Heart-shaped Watch Present Sale". made with a duplicated copy of the illustration setting, it was deliberately out of focus. Color separation was performed, then a duplicated copy with correct focus was added, superimposing only the line drawing part.

440
Ladies' Leather Boutique
Lario
Hong Kong
1983

AD／Alan Chan
D／Alan Chan, Alvin Chan
DF／Alan Chan Design Co.
"Lario" is Ladies' Leather boutique selling shoes and handbags imported from Europe. The shopping bag features a transparent tube handle that ties-in with Cinderella's glass slipper.

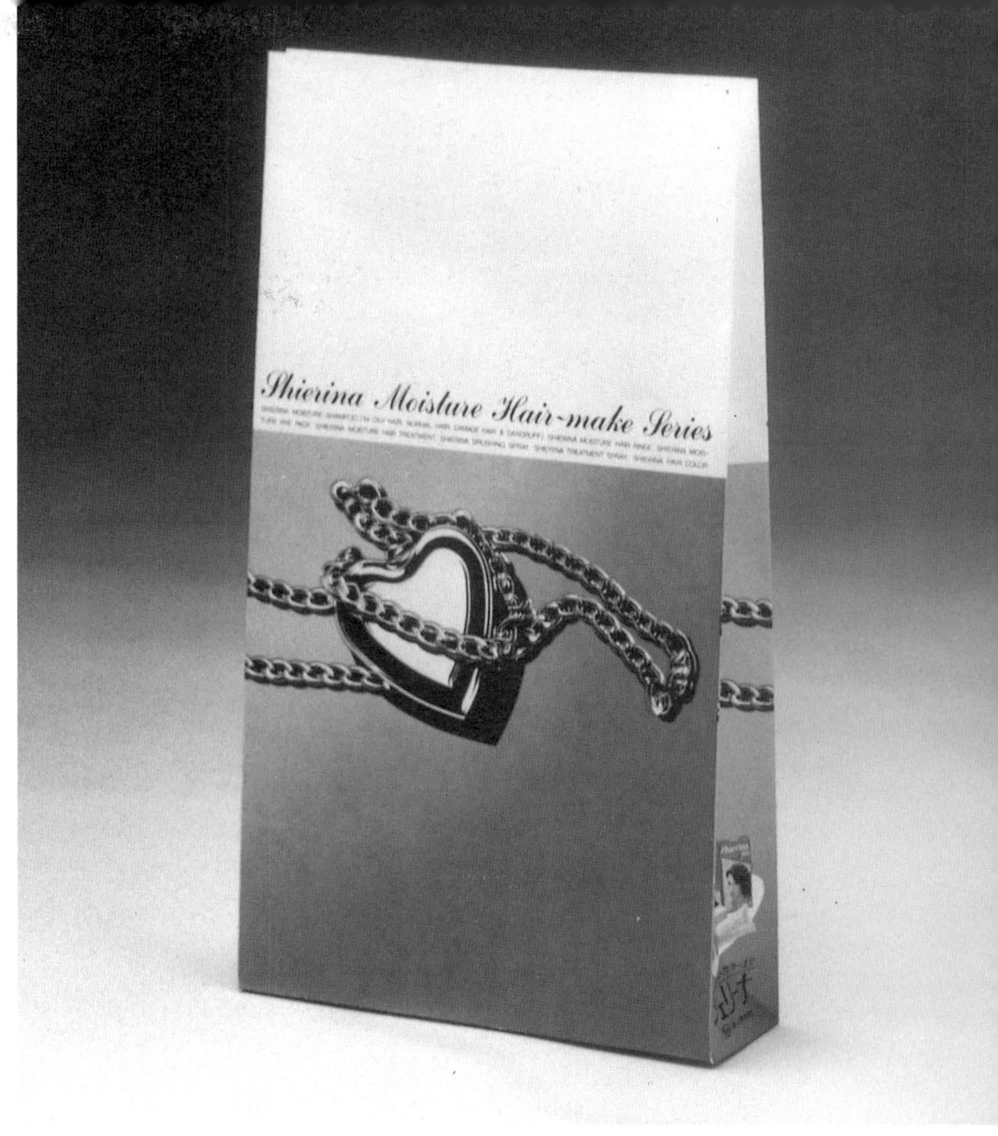

439

440

441

442

441・442
チャイニーズ・グッズ・ストア
ポリエチレン，紙
大中
大阪
1980—1987

D／野崎印刷紙業

441・442
Chinese Goods Store
Plastic, Paper
The Dai'chu, Inc.
Osaka, Japan
1980—1987

D／Nozaki Printing and Paper Industry Co., Ltd.

443
デパート催事用ショッピングバッグ
伊勢丹
東京
1987

AD／戸田正寿
D／鈴木 守
P／若月 勤
DF／戸田事務所，真木準企画室

443
Department Store Exhibition Shopping Bag
Isetan Department Store Co., Ltd.
Tokyo, Japan
1987

AD／Masatoshi Toda
D／Mamoru Suzuki
P／Tsutomu Wakatsuki
DF／Toda Studio, Jun Maki Planning Room

443

444
レストランのおみやげ用バッグ
モン・ジャルダン
東京
1984

D／小林純雄
I／ケート・グリンナウェイ，小池すみお
DF／マック・ジャパン
ケート・グリンナウェイのイラストをアイデンティティとして使用，女性向けレストランのイメージを打ち出している。このイラストは，マッチ，看板，名刺，コースターなどにも同じく使用される。

445
デパート催事用ショッピングバッグ
伊勢丹
東京
1987

AD／戸田正寿
D／鈴木 守
I／武田育雄
DF／戸田事務所，真木準企画室

446
美術大学
ビニール
嵯峨美術短期大学
京都
1987

D／鯛天成雄

444
Souvenir Paper Bag for Restaurant
Mon Jardin
Tokyo, Japan
1984

D／Sumio Kobayashi
I／Kate Greenaway, Sumio Koike
DF／Mac Japan Co., Ltd.
This design uses an illustration of Kate Greenaway as an image to express the idea of a restaurant for women. The same illustration is used for the restaurant's match book, sign board, business cards and coasters, too.

445
Department Store Exhibition Shopping Bag
Isetan Department Store Co., Ltd.
Tokyo, Japan
1987

AD／Masatoshi Toda
D／Mamoru Suzuki
I／Ikuo Takeda
DF／Toda Studio, Jun Maki Planning Room

446
Art College
Plastic
Saga Junior College of Art
Kyoto, Japan
1987

D／Nario Taiten

444

445

446

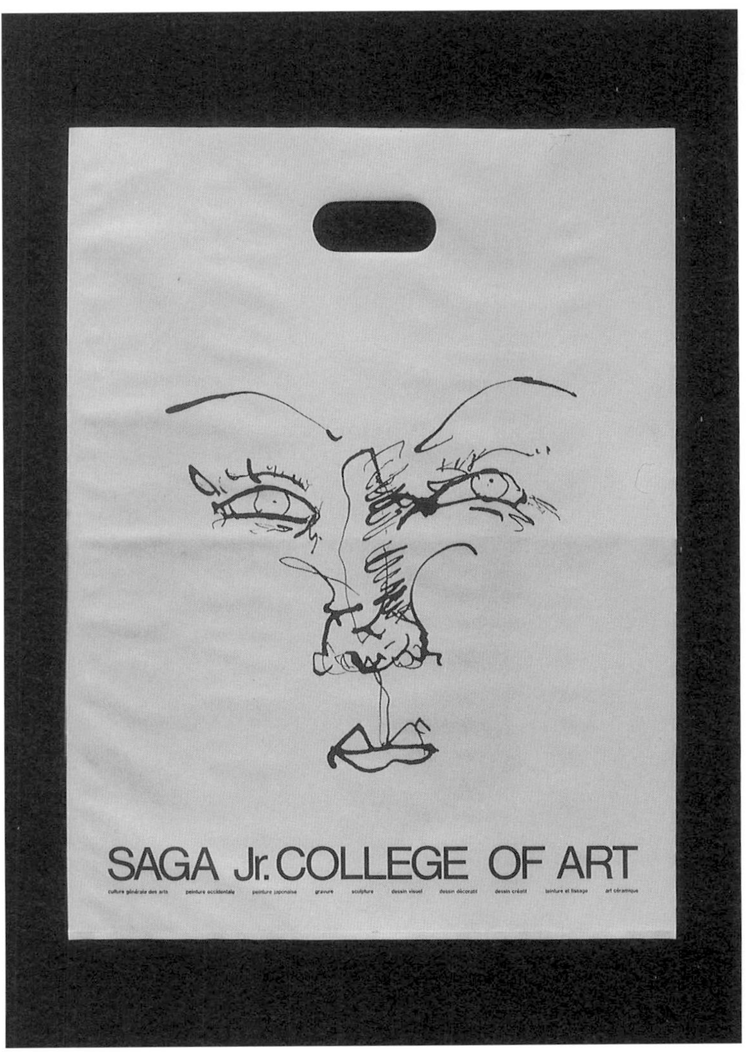

447

448

449

447
ファッション・ブランド"ジュン・メン"
ジュン
東京
1986

D／サイトウ・マコト，ジュン

448
アパレル・ブティック
ビニール
Zoobon
大阪
1977

D／川口紙業

449
メンズ・ブティック
パイル
大阪
1987

D／森口宏一
DF／尾崎紙工所
メンズショップのバッグ。ただし，女性にも好感を持たれるように素材を選択。素材はやや硬いシープスキンという紙を使っている。

450
タイヤ・ゴム製品会社
ブリヂストン
東京
1986

D/山本正典
DF/ブリヂストン・デザインセンター
コーポレート・カラーの赤・黒・白をベースに，社名ロゴをシルバーで表現し，使用目的に応じて3タイプを使い分ける。

451
スポーツ・ブティック
Chix
神戸市
1984

D/東京ステレオ・スタジオ

450

451

447
Fashion Brand "Jun Men"
Jun Co., Ltd.
Tokyo, Japan
1986

D/Makoto Saito, Jun Co., Ltd.

448
Apparel Boutique
Plastic
Zoobon
Osaka, Japan
1977

D/Kawaguchi Paper Industry Co., Ltd.

449
Men's Boutique
Pile Co., Ltd.
Osaka, Japan
1987

D/Hirokazu Moriguchi
DF/Ozaki Paper Products Co., Ltd.
A bag for a shop selling men's wear. The material, however, has been calculated to appeal tp women also : a rather stiff type of paper having the appearance of sheepskin.

450
Tire & Gum Company
Bridgestone Corporation
Tokyo, Japan
1986

D/Masanori Yamamoto
DF/Bridgestone Corporation
Using the corporate colors of black, red, and white in the background, the company logo is displayed in silver. The bag is available in three types, each appropriate for a different purpose.

451
Sporting Goods Store
Chix
Kobe City, Japan
1984

D/Tokyo Stereo Studio

453

452
アパレル・ブティック
ビニール
Gordini
大阪

D／Gordini

453
ファッション・ブランド"サルバドーレ・フェラガモ"
アオイ
東京
1987

D／サルバドーレ・フェラガモ(イタリア)
DF／下島包装開発

452
Apparel Boutique
Plastic
Gordini
Osaka, Japan

D／Gordini

453
Fashion Brand "Salvadore Ferragamo"
Aoi Co., Ltd.
Tokyo, Japan
1987

D／Salvadore Ferragamo (Italy)
DF／Shimojima Packaging Co., Ltd.

454

455

454
アパレル・ブティック
ビニール
ランバージャック
神戸市
1987

D／池永一彦
DF／尾崎紙工所

455
ブティック
ビニール
ホビーショップ・サンドウ
和歌山市
1986

I／貴田和恵
DF／ゴジラ企画
A／布谷

454
Apparel Boutique
Plastic
Lumber Jack
Kobe City, Japan
1987

D／Kazuhiko Ikenaga
DF／Ozaki Paper Products Co., Ltd.

455
Boutique
Plastic
Hobby Shop Sando
Wakayama City, Japan
1986

I／Kazue Kida
DF／Gozira Kikaku Co., Ltd.
A／Nunotani Co., Ltd.

国内出品者住所録

五十音順。住所の英語表記は，
出品者英語住所録欄に
アルファベット順で掲載。

㈱アイドル・メイクレット 〒151 東京都渋谷区千駄ヶ谷5-27-3 (03-352-5501)
㈱アオイ 〒102 東京都千代田区紀尾井町4-1 (03-239-0341)
㈱青山商品研究所 〒108 東京都港区芝4-11-5 MSビル5F (03-769-2678)
赤坂米穀㈱ 〒107 東京都港区南青山2-12-12 加藤ビル (03-401-2181)
浅葉克己 〒107 東京都港区南青山3-9-2 ㈱浅葉克己デザイン室 (03-479-0471)
ATTIC 〒650 兵庫県神戸市中央区三宮町2-10-7 (078-391-7692)
㈱アトリエ・ニキティキ 〒180 東京都武蔵野市吉祥寺本町2-28-3 (0422-21-8590)
ア・ストア・ロボット 〒150 東京都渋谷区恵比寿西2-17-16 (03-780-5825)
㈱アナザーワン 〒152 東京都目黒区自由ケ丘2-14-7 (03-717-2230)
Ami 〒150 東京都渋谷区宇田川町15-1 渋谷パルコパート3 1F (03-496-6103)
㈱Alicia ページボーイ事業部 〒150 東京都渋谷区神宮前4-9-3 (03-404-0643)
㈱アン 〒530 大阪市北区芝田町1-1-3 (06-373-0078)
五十嵐威暢 〒107 東京都港区南青山6-6-22 ㈱イガラシ・ステュディオ (03-498-3621)
池田武仁 〒790 愛媛県松山市朝生田町469 今居ビル1F Buzin Art (0899-45-0277)
石田 隆 〒462 愛知県名古屋市北区金城1-2-2-227 (052-914-7696)
市ロ清一 〒104 東京都中央区新富1-12-2 第二帝興ビル ㈱オーエムシー (03-553-6111)
㈱イッセイミヤケ・アンド・アソシエーツ 〒102 東京都千代田区五番町14-1 (03-265-0651)
㈱イッセイミヤケインターナショナル 〒102 東京都千代田区五番町14-1 (03-265-0655)
伊藤 満 〒104 東京都中央区銀座7-5-5 ㈱資生堂宣伝部 (03-572-5111)
㈱INAX 〒104 東京都中央区京橋3-6-18 (03-561-1710)
今村倶子 〒103 東京都中央区日本橋3-6-2 コーセー化粧品 (03-273-1518)
㈱イメージ・イン・スタジオ 〒060 札幌市中央区大通西16丁目 池川ビル (011-611-1230)
㈱岩田屋販売促進部 〒810 福岡市中央区天神2-11-1 (092-721-1111)
㈱イング・セブン 〒150 東京都渋谷区神宮前5-46-3 (03-498-3971)
㈲イングデザイン研究所 〒815 福岡市南区高宮4-11-32 (092-531-7234)
インタークルー 〒466 愛知県名古屋市昭和区壇渓通5-6 リバーサイドテラス石川橋 (052-834-6253)
ウィークエンズ 〒150 東京都渋谷区神宮前1-8-21 アトリウム85 (03-401-8350)
宇田 優 〒556 大阪市浪速区元町1-4-22-602 ユウ・デザイン事務所 (06-647-3711)
宇宙百貨 〒150 東京都渋谷区宇田川町4-9 くれたけビル1F (03-464-4148)
Wooden Doll 〒150 東京都渋谷区神宮前1-6-6 (03-423-0417)
㈱ヴァーナル・アンド・カンパニー 〒102 東京都千代田区一番町6-1 ロイヤル一番町707 (03-234-7533)
Voxel Yamamoto 〒604 京都市中京区河原町三条下ル大黒町55 (075-221-0969)
㈱エージー 〒104 東京都中央区銀座1-4-9 第一田村ビル (03-348-4483)
太田雄二 〒100 東京都千代田区丸の内1-1-1 パレスビル9F ㈱サン・アド (03-274-5021)
大西洋介 〒106 東京都港区西麻布1-14-15 ユーキ・フラット3B (03-408-0841)
岡田宏三 〒106 東京都港区六本木7-12-23 六本木フォルトゥナ2-A (03-408-8664)
奥村昭夫 〒530 大阪市北区西天満5-13-11 大東ビル2F ㈱パッケージング・クリエイト (06-312-7978)
㈱尾崎紙工所 〒558 大阪市東住吉区桑津1-15-15 (06-719-2967)
㈱オゾン・コミュニティー 〒151 東京都渋谷区千駄谷3-12-6 (03-478-8471)
㈱柿安本店 〒511 三重県桑名市江戸町36 (0594-23-5500)
片岡 脩 〒150 東京都渋谷区神宮前1-14-32 相互住宅原宿アパート203 ㈱シウ・グラフィカ (03-405-6755)
加藤多恵子 〒460 愛知県名古屋市中区丸の内3-18-12 ナカガワビル1A (052-961-9570)
上條喬久 〒107 東京都港区南青山7-8-1 小田急南青山ビル6F ㈱上條スタジオ (03-406-8641)
カルロス・ファルチ・ジャパン㈱ 〒106 東京都港区六本木7-18-5-109 (03-423-0169)
㈱鹿六 メディア・ショップ 〒604 京都市中京区河原町三条下ル一筋目東入ル大黒町44 (075-255-0081)
河北秀也 〒104 東京都中央区銀座1-16-3 篠原ビル ㈱日本ベリエール・アートセンター (03-564-1501)
川崎修司 〒150 東京都渋谷区恵比寿南1-13-2 エビスコート403 ㈲ヘッズ (03-713-2761)
ガストハウス 〒180 東京都武蔵野市吉祥寺東町1-6-16 (0422-21-8337)
貴田和恵 〒552 大阪市港区波除5-6-10 マンハイム弁天町604 (06-582-4629)
木田安彦 〒604 京都市中京区高倉通錦小路上ル 四条高倉スカイハイツ1103 (075-223-1847)
㈲Kid's Monster 〒150 東京都渋谷区恵比寿西1-34-15 (03-463-5425)
㈱キデイランド 〒150 東京都渋谷区神宮前5-7-20 神宮前太田ビル6F (03-409-3434)
㈱キムラタン 〒650 兵庫県神戸市中央区加納町2-9-24 (078-241-4351)
㈱Can 〒166 東京都杉並区高円寺北2-7-4 (03-339-1731)
㈱銀座千疋屋 〒104 東京都中央区銀座8-8-8 (03-572-0101)
銀座博品館 〒104 東京都中央区銀座8-8-11 (03-571-8008)
草刈 順 〒154 東京都世田谷区駒沢4-10-8 (03-421-3946)
熊谷英博 〒184 東京都小金井市桜町2-1-23 小金井コーポラス2-407 (0423-85-2067)
㈲クラブ・ジャップス 〒651 兵庫県神戸市中央区中島通1-5-23 (078-221-1291)
グラマー 〒542 大阪市南区南炭屋町30-1 (06-211-2534)
グランド・キャニオン㈱ 〒150 東京都渋谷区猿楽町9-8 (03-463-1161)
㈱グレート・ブリテン 〒115 東京都北区赤羽1-1-7 (03-903-5777)
㈱ケンウッド，コーポレート・デザイン部 〒150 東京都渋谷区渋谷2-17-5 シオノギ渋谷ビル (03-486-5645)
KENZO Paris 〒107 東京都港区南青山6-7-5-710 (03-407-3271)
㈲現代衣学 〒150 東京都渋谷区神宮前6-35-3 コープ・オリンピア322 (03-797-0754)
小島 歩 〒462 愛知県名古屋市北区東水切町4-48 (052-981-3256)
コンガ 〒150 東京都渋谷区代官山町20-6 (03-477-2620)
Gordini 〒542 大阪市南区心斎橋2-20 (06-213-3474)
斎藤喜郎 〒020 岩手県盛岡市開運橋通3-43 マンション菜園205 ㈲ハイプロダクション (0196-52-1519)
榊原勝一 〒102 東京都千代田区六番町4 朝日六番町マンション405 デザイン・スタジオ202 (03-221-9481)
佐古田英一 〒665 兵庫県宝塚市平井山荘20-5 (0797-89-2257)
㈱サープラス 〒150 東京都渋谷区神宮前5-28-7 グリーンタウン306 (03-400-5049)
沢 正一 〒104 東京都中央区銀座6-13-3 ㈱マグナ (03-543-4341)
㈱サンシャインシティ，コミュニケーション部 〒170 東京都豊島区東池袋3-1 (03-989-3531)
㈱サン・デザイン・アソシエーツ 〒541 大阪市東区南久太郎町1-29 大成ビル7F (06-261-2961)
サントリー㈱ デザイン部 〒530 大阪市北区堂島浜2-1-40 (06-346-1140)

三洋電機㈱ 販売促進課　〒570 大阪府守口市大日東町100（06-901-1111）
㈱サンリオ　〒141 東京都品川区大崎1-6-1（03-779-8111）
㈱ザ・ギンザ商品企画室　〒104 東京都中央区銀座7-4-12 行政ビル5F（03-572-2121）
㈱ザ・デザイン・アソシエイツ　〒106 東京都港区麻布台2-3-8 丸山ビル（03-587-1740）
㈱シェトワ　〒542 大阪市南区谷町9-5-15 中田ビル4F（06-762-9521）
㈱C カンパニー　〒102 東京都千代田区一番町15 壱番館ビル3F（03-230-4588）
シティ・ユニオン　〒107 東京都港区南青山5-4-44 ラポール南青山B101（03-499-6034）
嶋 高宏　〒550 大阪市西区江戸堀1-22 大同生命ビル8F ㈱嶋デザイン事務所（06-441-8188）
シモジマ商事㈱　〒111 東京都台東区浅草橋5-29-8（03-861-6059）
ヘルムート・シュミット　〒564 大阪府吹田市垂水町3-24-14-707（06-338-5566）
白川こう吉　〒370 群馬県高崎市昭和町142 グリーンハイツ300 白川こう吉デザイン室（0273-62-1318）
㈱シン＆カンパニー　〒141 東京都品川区西五反田5-13-1 フランス屋ビル（03-495-6691）
㈱シンシア　〒156 東京都世田谷区船橋2-7-6（03-789-2281）
㈱新宿高野　〒160 東京都新宿区新宿3-26-11（03-354-0222）
㈱ジーピー　〒333 埼玉県川口市上青木西5-23-28（0482-66-0802）
㈱ジュン　〒108 東京都港区港南1-8-22（03-472-4111）
㈱ジョイント　〒564 大阪府吹田市豊津9-1 東洋ビル（06-380-3405）
鈴木善博　〒104 東京都中央区銀座3-12-14 秋葉ビル3F ㈱ビーバイスタジオ（03-546-0918）
鈴木 勝　〒462 愛知県名古屋市北区辻町1-8 辻町住宅2-3-105（052-915-1387）
Zoobon　〒542 大阪市南区東清水町35-9 水谷ビル1F（06-241-1808）
ゼロワン原宿店　〒150 東京都渋谷区神宮前6-10-10 原ビル1F（03-486-3601）
㈱ソニー・クリエイティブ・プロダクツ　〒102 東京都千代田区紀尾井町3-6（03-237-5645）
㈱ソニープラザ　〒104 東京都中央区銀座6-8-5 小松アネックスビル（03-575-2511）
鯛天成雄　〒601 京都市南区吉祥院石原長田町1-1 桂川ハイツ5-211 T-Box（075-661-2960）
㈱高島屋　〒103 東京都中央区日本橋2-4-1（03-211-4111）
高田雄吉　〒541 大阪市東区北浜1-33 ㈱アイ・エフ・プランニング（06-231-8282）
高橋 篤　〒658 兵庫県神戸市東灘区御影本町6-11-19 モロゾフ㈱ マーケティング本部第１課デザイン室（078-822-5032）
竹内二朗　〒060 札幌市中央区北4条西15丁目 ㈲ミッキーハウス（011-644-2543）
武下 朗　〒107 東京都港区赤坂7-6-41-308 ㈱武下CD事務所（03-589-0913）
竹智 淳　〒160 東京都新宿区新宿1-26-12-1001 ㈲バーヴ（03-341-3649）
立松 脩　〒460 愛知県名古屋市中区丸の内3-19-1 ライオンビル6F ㈱デコム・クリエーティブアート＆サービス（052-951-6591）
田中康夫　〒558 大阪市住吉区帝塚山中1-3-2 帝塚山タワープラザ201 ㈱パッケージランド（06-675-0138）
㈱大中　〒564 大阪府吹田市豊津町9-1 ダイエー江坂オフィスセンタービル16F（06-380-4531）
㈱大和シュガル　〒150 東京都渋谷区恵比寿西2-16-13 代官山パンション303（03-461-8367）
Chix　〒650 兵庫県神戸市中央区北野町4-1-12（078-222-4810）
チチカカ　〒150 東京都渋谷区宇田川町6-20 パラシオン渋谷104（03-476-3449）
チャンプ・カンパニー　〒650 兵庫県神戸市中央区三宮町1-8-1 サンプラザビル6F（078-331-9544）
塚本佳哉樹　〒106 東京都港区六本木5-1-4 六和ビル5F コミュニケーションアーツR（03-405-0270）
TDK㈱ 磁気テープ事業部企画部宣伝販促課　〒103 東京都中央区日本橋1-13-1（03-278-5043）
㈱ディ・ショップ　〒150 東京都渋谷区神宮前1-9-11（03-478-3941）
ディボア　〒150 東京都渋谷区恵比寿西1-33-18 コート代官やま102（03-463-6145）
デザインオフィス・ルームワン　〒462 愛知県名古屋市北区大蔵町53 マンションオークラ1-D（052-915-7547）
デライツ・クラブ　〒150 東京都渋谷区神宮前1-17-5 原宿シュロス402（03-470-3567）
戸田正寿　〒104 東京都中央区築地7-16-3 クラウン築地307 戸田事務所（03-545-1533）
㈱トライ　〒151 東京都渋谷区千駄ヶ谷3-4-10 原宿葵マンション（03-401-4337）
鳥山大樹　〒530 大阪市北区与力町6-6-205 バード・デザインハウス（06-354-1797）
㈱ドット　〒150 東京都渋谷区渋谷2-7-6（03-407-3150）
㈱ナイス　〒154 東京都世田谷区若林4-9-15 大研ビル1F（03-487-0045）
㈱ナイス・クリエーション　〒531 大阪市大淀区大淀中2-7-6（06-453-2851）
NYLON　〒532 大阪市淀川区宮原3-3-36（06-391-4951）
中崎宣弘　〒107 東京都港区元赤坂1-2-3 サントリー㈱ 東京支社パッケージング部（03-470-9827）
中谷匡児　〒151 東京都渋谷区千駄ヶ谷3-16-3 メイゾン原宿401 中谷匡児デザイン室（03-402-2549）
中村政久　〒150 東京都渋谷区広尾4-1-14-1307（03-797-6137）
永井一正　〒104 東京都中央区銀座1-13-13 ㈱日本デザインセンター（03-567-3231）
長友啓典　〒106 東京都港区六本木7-18-7 内海ビル4F K2（03-401-9266）
成瀬政敏　〒573 大阪府枚方市楠葉朝日2-18-15 パノラマ・アドバタイジング（0720-56-2246）
成瀬始子　〒107 東京都港区南青山2-7-6 成瀬始子デザイン室（03-401-4481）
野口正治　〒150 東京都渋谷区神宮前1-21-1-302（03-405-6814）
ノース・マリン・ドライブ　〒107 東京都港区南青山7-4-17 南青山131ビル103（03-406-8045）
野村たかあき　〒371 群馬県前橋市朝日町4-1-18（0272-43-7061）
㈱ハニー　〒150 東京都渋谷区恵比寿南1-1-1（03-715-7161）
早川良雄　〒107 東京都港区赤坂8-5-28 コープ赤坂ハイツ302 早川良雄デザイン事務所（03-475-1051）
林 隆史　〒456 愛知県名古屋市熱田区金山町1-7-5 住友生命金山第２ビル2F ㈱サンライト（052-681-3351）
原宿サーキット　〒150 東京都渋谷区神宮前1-8-8 COXY188 1F（03-404-8993）
原田 治　〒104 東京都中央区明石町13-17-802（03-542-7704）
㈱バルーン　〒152 東京都目黒区自由ヶ丘1-26-13（03-717-4144）
パウンド・ハウス　〒542 大阪市南区大宝寺町中之丁29（06-251-4334）
㈱パピルス　〒150 東京都渋谷区神宮前2-24-4 里見ビル1F（03-402-5401）
Beat Pops by Jemmy's　〒150 東京都渋谷区神宮前1-15-1 ビア原宿1F（03-478-6059）
B×3㈱ NODE館　〒150 東京都渋谷区神宮前6-12-20（03-499-2132）
㈱ビームス　〒150 東京都渋谷区神宮前3-21-8（03-470-9391）
ピーナッツ・ボーイ　〒150 東京都渋谷区神宮前4-26-24（03-479-0780）
ピンク・バス　〒150 東京都渋谷区宇田川町13（03-770-2070）
45RPM-Studio Co.　〒107 東京都港区南青山7-7-21（03-486-0045）
藤重日生　〒733 広島市中区堺町2-4-21 サンピア白樺407（082-292-6150）
舟橋全二　〒249 神奈川県逗子市逗子6-5-20（0468-73-7328）

㈱フラミンゴ・スタジオ　〒160 東京都新宿区新宿6-3-11 (03-352-9717)
㈱フランドル　〒151 東京都渋谷区千駄ヶ谷2-9-6 (03-470-6141)
㈱フリーウェイ　〒152 東京都目黒区自由ヶ丘2-16-10 (03-723-2632)
Free Hand　〒150 東京都渋谷区神宮前1-8-8 COXY188 1F (03-405-4535)
古田 修　〒150 東京都渋谷区神宮前3-38-12 原宿ロイヤル8F-E (03-402-1001)
㈱フロンティア　〒150 東京都渋谷区神南1-16-3 (03-464-4579)
ブルー＆ホワイト　〒106 東京都港区麻布十番2-9-2 (03-451-0537)
㈱ブルーグラス　〒103 東京都中央区日本橋本町1-6-1 (03-246-8865)
㈲プラネット　〒111 東京都台東区柳橋2-16-14 (03-865-0991)
㈱プランタン銀座　〒104 東京都中央区銀座3-2-1 (03-567-0077)
㈱プロライン　〒107 東京都港区赤坂3-21-17 NTCビル (03-587-0271)
㈱ペーパー・ムーン・インターナショナル・ジャパン　〒150 東京都渋谷区神宮前2-31-7 ビラ・グロリア802 (03-478-4320)
ホームズ・アンダーウェア　〒150 東京都渋谷区恵比寿西1-21-4 (03-464-7072)
㈱ポップショップ・エンタープライズ　〒107 東京都港区南青山3-11-12 (03-408-0207)
ポルトフィーノ　〒530 大阪市北区芝田1-1-3 阪急三番街 (06-374-0009)
㈱マック・ジャパン　〒105 東京都港区虎ノ門3-25-3 芝ロイヤル501 (03-436-3676)
㈱松坂屋　〒104 東京都中央区銀座6-10-1 (03-572-1111)
松永 真　〒107 東京都港区南青山7-3-1 石橋興業ビル8F ㈱松永真デザイン事務所 (03-499-0291)
㈱三越 デザイン開発室　〒103 東京都中央区日本橋室町1-4 (03-274-7014)
㈱光村原色版印刷所 制作室　〒141 東京都品川区大崎1-15-9 (03-492-1190)
村井和章　〒104 東京都中央区銀座7-5-5 ㈱資生堂宣伝部 (03-572-5111)
村越幸子　〒150 東京都渋谷区東1-4-1 尚豊ビル606 ㈱ジャイア (03-400-4210)
村越 襄　〒106 東京都港区麻布台1-1-20 麻布台ユニ・ハウス514 ㈱村越襄デザイン室 (03-584-5810)
㈱メンズショップ・トラヤ　〒542 大阪市南区難波1-5-14 (06-211-1522)
㈱メンズショップ・トラヤ Lamp Fox　〒661 兵庫県尼崎市塚口本町4-8-1 つかしんモール街M-155 (06-420-3663)
森 貞人　〒461 愛知県名古屋市東区泉1-22-35 チサンマンション桜通久屋304 (052-961-7508)
㈱森久第二事務所　〒542 大阪市南区南炭屋町11 スパジオ6F (06-211-8951)
八尾武郎　〒160 東京都新宿区本塩町9-3 司法書士会館3F ㈱YAOデザイン研究所 (03-357-3686)
八木健夫　〒150 東京都渋谷区渋谷1-3-18 ビラモデルナA406 ㈱オフィース・ピーアンドシー (03-406-0191)
安原和夫　〒104 東京都中央区銀座7-5-5 ㈱資生堂宣伝部 (03-572-5111)
矢萩喜従郎　〒151 東京都渋谷区本町1-39-1-203 (03-375-9204)
山内瞬葉　〒464 愛知県名古屋市千種区四谷通3-26 四谷ビル305 山内瞬葉デザイン室 (052-782-6989)
山崎達雄　〒103 東京都中央区日本橋3-9-2 第二丸善ビル 丸善㈱ 宣伝制作課 (03-272-7033)
㈱やまもと寛斎　〒150 東京都渋谷区神宮前4-3-15 東京セントラル表参道ビル (03-470-2030)
山本正典　〒106 東京都港区六本木5-17-1 AXISビル5F ㈱ブリヂストン デザインセンター (03-583-0311)
USPP　〒151 東京都渋谷区千駄ヶ谷3-41-14 (03-478-4120)
㈱横浜岡田屋(岡田屋モアーズ)　〒220 神奈川県横浜市西区南幸1-3-1 (045-311-1471)
㈱ライカ 広報宣伝課　〒534 大阪市都島区都島北通2-8-17 (06-922-8423)
ラパス　〒650 兵庫県神戸市中央区三宮町1-8-1 サンプラザビル6F (078-332-3707)
㈱ラフォーレ原宿　〒150 東京都渋谷区神宮前6-2-9 (03-486-9881)
ランバージャック　〒650 兵庫県神戸市中央区三宮町1-8-1 サンプラザ6F (078-392-3700)
㈱リチャード　〒542 大阪市南区心斎橋筋2-31 (06-211-4477)
㈱リトルアンデルセン ヒステリックミニ事業部　〒150 東京都渋谷区神宮前5-41-5 (03-486-8431)
㈱ル・ギャマン　〒150 東京都渋谷区神宮前3-31-20 (03-402-5204)
ロス　〒650 兵庫県神戸市中央区北長狭通1-31-35 (078-321-3394)
㈱ロック座　〒101 東京都千代田区神田小川町2-1 (03-295-1545)
Y. Inohana & Sons　〒604 京都市中京区新京極三条下ル (075-221-3561)
㈱若菜　〒170 東京都豊島区東池袋3-1-1 サンシャイン60 43F (03-989-5962)
㈱ワシントン靴店　〒104 東京都中央区銀座5-7-7 (03-572-5911)
㈱わちふぃーるど　〒152 東京都目黒区自由ヶ丘2-19-5 (03-725-0881)
㈱ワールド　〒650 兵庫県神戸市中央区港島中町6-8-1 (078-302-3111)
ワンダーハウス　〒150 東京都渋谷区宇田川町11-2 (03-461-6266)

出品者英語住所録
Contributors Addresses

[Japan]

A Store Robot 2-17-16 Ebisu-nishi Shibuya-ku, Tokyo 150, Japan (03-780-5825)
A.G.Co., Ltd. Dai-ichi Tamura Bldg., 1-4-9 Ginza Chuo-ku, Tokyo 104, Japan (03-348-4483)
Akasaka Beikoku Co., Ltd. Kato Bldg., 2-12-12 Minami-Aoyama Minato-ku, Tokyo 107, Japan (03-401-2181)
Alicia Co., Ltd. 4-9-3 Jingumae Shibuya-ku, Tokyo 150, Japan (03-404-0643)
Ami 1F Shibuya Parco Part III, 15-1 Udagawa-cho Shibuya-ku, Tokyo 150, Japan (03-496-6103)
Ann Co., Ltd. 1-1-3 Shibata-cho Kita-ku, Osaka 530, Japan (06-373-0078)
Another One Co., Ltd. 2-14-7 Jiyugaoka Meguro-ku, Tokyo 152, Japan (03-717-2230)
Aoi Co., Ltd. 4-1 Kinoi-cho Chiyoda-ku, Tokyo 102, Japan (03-239-0341)
Aoyama Research Institute of Fashion Merchandising Co., Ltd. MS Bldg., 4-11-5 Shiba Minato-ku, Tokyo 108, Japan (03-769-2678)
Katsumi Asaba Asaba Design Co., Ltd. 3-9-2 Minami-Aoyama Minato-ku Tokyo 107, Japan (03-479-0471)
Atelier Niki Tiki Co., Ltd. 2-28-3 Kichijoji-honmachi Musashino City, Tokyo 180, Japan (0422-21-8590)
ATTIC 2-10-7 Sannomiya-cho Chuo-ku Kobe City, Hyogo 650, Japan (078-391-7692)
Balloon Co., Ltd. 1-26-13 Jiyugaoka Meguro-ku, Tokyo 152 (03-717-4144)
B-By-Three Co., Ltd. 6-12-20 Jingumae Shibuya-ku, Tokyo 150, Japan (03-499-2132)
Beams Co., Ltd. 3-12-8 Jingumae Shibuya-ku, Tokyo 150, Japan (03-470-9391)
Beat Pops by Jemmy's 1F Via Harajuku, 1-15-1 Jingumae Shibuya-ku, Tokyo 150, (03-478-6059)
Blue & White 2-9-2 Azabu-Juban Minato-ku, Tokyo 106, Japan (03-451-0537)
Blue Glass Co., Ltd. 1-6-1 Nihonbashi-honcho Chuo-ku, Tokyo 103, Japan (03-246-8865)
C Company Limited 3F Ichibankan Bldg., 15 Ichiban-cho Chiyoda-ku, Tokyo 102, Japan (03-230-4588)
Can Company Ltd. 2-7-4 Koenji-kita Suginami-ku, Tokyo 166, Japan (03-339-1731)
Carlos Falchi Japan Co., Ltd. 7-18-5-109 Roppongi Minato-ku, Tokyo 106, Japan (03-423-0169)
Champ Company 6F Sun Plaza Bldg., 1-8-1 Sannomiya-cho Chuo-ku Kobe City, Hyogo 650, Japan (078-331-9544)
Chez Toi Co., Ltd. 4F Nakata Bldg., 9-5-15 Tani-machi Minami-ku, Osaka 542 (06-762-9521)
Chix 4-1-12 Kitano-cho Chuo-ku Kobe City, Hyogo 650, Japan (078-222-4810)
Club Japs Inc. 1-5-23 Nakajima-dori Chuo-ku Kobe City, Hyogo 651, Japan (078-221-1291)
City Union B101 Rapport Minami-Aoyama, 5-4-44 Minami-Aoyama Minato-ku, Tokyo 107, Japan (03-499-6034)
Cynthia Corp. Ltd. 2-7-6 Funabashi Setagaya-ku, Tokyo 156, Japan (03-789-2281)
D. Shop Co., Ltd. 1-9-11 Jingumae Shibuya-ku, Tokyo 150, Japan (03-478-3941)
The Dai'chu, Inc. 16F Daie Esaka Office Center Bldg., 9-1 Toyotsu-cho Suita City, Osaka 564, Japan (06-380-4531)
D'Lites Club 1-17-5-402 Jingumae Shibuya-ku, Tokyo 150, Japan (03-470-3567)
Design Office Room-One 1-D Mansion Okura, 53 Okura-cho Kita-ku, Nagoya 462, Japan (052-915-7547)
Diboa 102 Court Daikanyama, 1-33-18 Ebisu-nishi Shibuya-ku, Tokyo 150, Japan (03-463-6145)
Dot Co., Ltd. 2-7-6 Shibuya, Shibuya-ku, Tokyo 150, Japan (03-407-3150)
Flamingo Studio Inc. 6-3-11 Shinjuku, Shinjuku-ku, Tokyo 160, Japan (03-352-9717)
Flandre Co., Ltd. 2-9-6 Sendagaya Shibuya-ku, Tokyo 151, Japan (03-470-6141)
45RPM-Studio Co. 7-7-21 Minami-Aoyamo Minato-ku, Tokyo 107, Japan (03-486-0045)
Free Hand 1F COXY 188, 1-8-8 Jingumae Shibuya-ku, Tokyo 150, Japan (03-405-4535)
Freeway Co., Ltd. 2-16-10 Jiyugaoka Meguro-ku, Tokyo 152, Japan (03-723-2632)
Frontier Co., Ltd. 1-16-3 Jinnan Shibuya-ku, Tokyo 150, Japan (03-464-4579)
Teruo Fujishige 407 Sanpia Shirakaba, 2-4-21 Sakai-machi Naka-ku, Hiroshima 733, Japan (082-292-6150)
Zenji Funabashi 6-5-20 Zushi, Zushi City, Kanagawa 249, Japan (0468-73-7328)
Osamu Furuta 8F-E Harajuku Royal, 3-38-12 Jingumae Shibuya-ku, Tokyo 150, Japan (03-402-1001)
G & P Co., Ltd. 5-23-28 Kami-Aoki-nishi Kawaguchi City, Saitama 333, Japan (0482-66-0802)
Gast Haus 1-6-16 Kichijoji-higashi Musashino City, Tokyo 180, Japan (0422-21-8337)
Gendai-Igaku Co., Ltd. 322 Coop Olympia, 6-35-3 Jingumae Shibuya-ku, Tokyo 150, Japan (03-797-0754)
Ginza Hakuhinkan 8-8-11 Ginza Chuo-ku, Tokyo 104, Japan (03-571-8008)
Ginza Senbikiya Co., Ltd. 8-8-8 Ginza Chuo-ku, Tokyo 104, Japan (03-572-0101)
Glamour 30-1 Minami-Sumiyamachi Minami-ku, Osaka 542, Japan (06-211-2534)
Gordini 2-20 Shinsaibashi Minami-ku, Osaka 542, Japan (06-213-3474)
Grand Canyon Co., Ltd. 9-8 Sarugaku-cho Shibuya-ku, Tokyo 150, Japan (03-463-1161)
Great Britain Corporation 1-1-7 Akabane Kita-ku, Tokyo 115, Japan (03-903-5777)
Osamu Harada 13-17-802 Akashi-cho Chuo-ku, Tokyo 104, Japan (03-542-7704)
Harajuku Circuit 1F COXY 188, 1-8-8 Jingumae Shibuya-ku, Tokyo 150, Japan (03-404-8993)
Yoshio Hayakawa Hayakawa Design Office, 302 Co-op Akasaka Heights, 8-5-28 Akasaka Minato-ku, Tokyo 107, Japan (03-475-1051)
Takafumi Hayashi Sunlight Co., Ltd., 2F Sumitomo-seimei Kanayama Dai-ni Building, 1-7-5 Kanayama-cho Atsuta-ku, Nagoya 456, Japan (052-681-3351)
Homes' Underwear 1-21-4 Ebisu-nishi Shibuya-ku, Tokyo 150, Japan (03-464-7072)
Honey Co., Ltd. 1-1-1 Ebisu-minami Shibuya-ku, Tokyo 150, Japan (03-715-7161)
Seiichi Ichiguchi OMC, Inc., Dai-ni Teikyo Bldg., 1-12-2 Shintomi Chuo-ku, Tokyo 104, Japan (03-553-6111)
Idol Makelet Co., Ltd. 5-27-3 Sendagaya Shibuya-ku, Tokyo 151, Japan (03-352-5501)
Takenobu Igarashi Igarashi Studio, 6-6-22 Minami-Aoyama Minato-ku, Tokyo 107, Japan (03-498-3621)
Takehito Ikeda Buzin Art, 1F Imai Bldg., 469 Asoda-machi Matsuyama City, Ehime 790, Japan (0899-45-0277)
Image Inn Studio Co., Ltd. Ikekawa Bldg., West-16-chome Odori Chuo-ku, Sapporo 060, Japan (011-611-1230)
Tomoko Imamura Kose Cosmetics Co., Ltd., 3-6-2 Nihonbashi Chuo-ku, Tokyo 103, Japan (03-273-1518)
INAX Corporation 3-6-18 Kyobashi Chuo-ku, Tokyo 104, Japan (03-561-1710)
. . . ing Design Institute 4-11-32 Takamiya Minami-ku, Fukuoka City 815, Japan (092-531-7234)
Ing Seven Co., Ltd. 5-46-3 Jingumae Shibuya-ku, Tokyo 150, Japan (03-498-3971)
Intercrew Riverside Terrace Ishikawabashi, 5-6 Dankei-dori Showa-ku, Nagoya 466, Japan (052-834-6253)
Takashi Ishida 1-2-2-227 Kinjo Kita-ku, Nagoya 462, Japan (052-914-7696)
Issey Miyake & Associates Inc. 14-1 Goban-cho Chiyoda-ku, Tokyo 102, Japan (03-265-0651)
Issey Miyake International Inc. 14-1 Goban-cho Chiyoda-ku, Tokyo 102, Japan (03-265-0655)
Mitsuru Ito Advertising Dept., Shiseido Co., Ltd., 7-5-5 Ginza Chuo-ku, Tokyo 104, Japan (03-572-5111)
Iwataya Department Store Co., Ltd. 2-11-1 Tenjin Chuo-ku, Fukuoka 810, Japan (092-721-1111)
Joint Co., Ltd. Toyo Bldg., 9-1 Toyotsu Suita City, Osaka 564, Japan (06-380-3405)
Jun Co., Ltd. 1-8-22 Konan Minato-ku, Tokyo 108, Japan (03-472-4111)
Kakiyasu-Honten 36 Edo-machi Kuwana City, Mie 511, Japan (0594-23-5500)
Takahisa Kamijyo Kamijyo Studio, 6F Odakyu Minami-Aoyama Bldg., 7-8-1 Minami-Aoyama Minato-ku, Tokyo 107, Japan (03-406-8641)
Karoku Co., Ltd. Media Shop 44 Daikoku-cho Kawaramachi-Sanjo Nakagyo-ku, Kyoto 604, Japan (075-255-0081)
Shu Kataoka Siu Graphica Inc., 203 Sogojutaku-Harajuku Apartment, 1-14-32 Jingumae Shibuya-ku, Tokyo 150, Japan

(03-405-6755)
Taeko Kato 1A Nakagawa Bldg., 3-18-12 Marunouchi Naka-ku, Nagoya 460, Japan (052-961-9570)
Hideya Kawakita Japan Bélier Art Center Inc., Shinohara Bldg., 1-16-3 Ginza Chuo-ku, Tokyo 104, Japan (03-564-1501)
Shuji Kawasaki Heads Inc., 403 Ebis Court, 1-13-2 Ebisu-minami Shibuya-ku, Tokyo 150, Japan (03-713-2761)
Kenwood Corporation Shionogi Shibuya Bldg., 2-17-5 Shibuya, Shibuya-ku, Tokyo 150, Japan (03-486-5645)
Kenzo Paris 6-7-5-710 Minami-Aoyama Minato-ku, Tokyo 107, Japan (03-407-3271)
Kazue Kida 604 Mannheim Bentencho, 5-6-10 Namiyoke Minato-ku, Osaka 552, Japan (06-589-4629)
Yasuhiko Kida 1103 Shijo-takakura Sky Heights, Takakura-dori Nishikikoji Agaru, Nakagyo-ku, Kyoto 604, Japan (075-223-1847)
Kiddy Land Co., Ltd. 6F Jingumae-Ota Bldg., 5-7-20 Jingumae Shibuya-ku, Tokyo 150, Japan (03-409-3434)
Kid's Monster Inc. 1-34-15 Ebisu-nishi Shibuya-ku, Tokyo 150, Japan (03-463-5425)
Kimuratan Co., Ltd. 2-9-24 Kano-cho Chuo-ku Kobe City, Hyogo 650, Japan (078-241-4351)
Ayumu Kojima 4-48 Higashi-Mizukiri-cho Kita-ku, Nagoya 462, Japan (052-981-3256)
Konga 20-6 Daikanyama-cho Shibuya-ku, Tokyo 150, Japan (03-477-2620)
Hidehiro Kumagai 2-1-23-2-407 Sakura-machi Koganei-shi, Tokyo 184, Japan (0423-85-2067)
Jun Kusakari 4-10-8 Komazawa Setagaya-ku, Tokyo 154, Japan (03-421-3946)
Laforet Harajuku 6-2-9 Jingumae Shibuya-ku, Tokyo 150, Japan (03-486-9881)
Lapaz 6F Sun Plaza Bldg., 1-8-1 Sannomiya-cho Chuo-ku Kobe City, Hyogo 650, Japan (078-332-3707)
Le Gamin S.A. 3-31-20 Jingumae Shibuya-ku, Tokyo 150, Japan (03-402-5204)
Little Andersen Co., Ltd. Hysteric Mini, 5-41-5 Jingumae Shibuya-ku, Tokyo 150, Japan (03-486-8431)
Los 1-31-35 Kita-Nagasa-dori Chuo-ku Kobe City, Hyogo 650, Japan (078-321-3394)
Lumber Jack 6F Sun Plaza, 1-8-1 Sannomiya-cho Chuo-ku Kobe City, Hyogo 650, Japan (078-392-3700)
Mac Japan Co., Ltd. 501 Shiba Royal, 3-25-3 Toranomon Minato-ku, Tokyo 105, Japan (03-436-3676)
Shin Matsunaga Shin Matsunaga Design Inc., 8F Ishibashi Kogyo Bldg., 7-3-1 Minami-Aoyama, Minato-ku Tokyo 107, Japan (03-499-0291)
Matsuzakaya Co., Ltd. 6-10-1 Ginza Chuo-ku, Tokyo 104, Japan (03-572-1111)
Men's Shop Toraya 1-5-14 Namba Minami-ku, Osaka 542, Japan (06-211-1522)
Men's Shop Toraya Lamp Fox M-155 Tsukashin Mall, 4-8-1 Tsukaguchi-honcho Amagasaki City, Hyogo 661, Japan (06-420-3663)
Mitsukoshi Limited 1-4 Nihonbashi-Muro-machi Chuo-ku, Tokyo 103, Japan (03-274-7014)
Mitsumura Printing Co., Ltd. 1-15-9 Osaki Shinagawa-ku, Tokyo 141, Japan (03-492-1190)
Sadahito Mori 304 Chisan-mansion Sakuradori-hisaya, 1-22-35 Izumi Higashi-ku, Nagoya 461, Japan (052-961-7508)
Morihisa Second Co., Ltd. 6F Spazio, 11 Minami-Sumiyamachi Minami-ku, Osaka 542, Japan (06-211-8951)
Morozoff Limited 6-11-19 Mikage-honmachi Higashinada-ku, Kobe 658, Japan (078-822-5032)
Kazuaki Murai Advertising Dept., Shiseido Co., Ltd., 7-5-5 Ginza Chuo-ku, Tokyo 104, Japan (03-572-5111)
Jo Murakoshi Murakoshi Jo Design Shitsu, 514 Azabudai Uni-House, 1-1-20 Azabudai Minato-ku, Tokyo 106, Japan (03-584-5810)
Murakoshi Sachiko Jia Co., Ltd., 606 Shoho Building, 1-4-1 Higashi Shibuya-ku, Tokyo 150, Japan (03-400-4210)
Kazumasa Nagai Nippon Design Center Co., Ltd., 1-13-13 Ginza Chuo-ku, Tokyo 104, Japan (03-567-3231)
Keisuke Nagatomo K-Two Co., Ltd., 4F Utsumi Bldg., 7-18-7 Roppongi Minato-ku, Tokyo 106, Japan (03-401-9266)
Kyoji Nakatani Nakatani Kyoji Design Studio, 401 Maison Harajuku, 3-16-3 Sendagaya Shibuya-ku, Tokyo 151, Japan (03-402-2549)
Nobuhiro Nakazaki Packaging Dept., Tokyo Branch, Suntory Limited, 1-2-3 Moto-Akasaka Minato-ku, Tokyo 107, Japan (03-470-9827)
Masatoshi Naruse Panorama Advertising Inc., 2-18-15 Kuzuha-Asahi Hirakata City, Osaka 573, Japan (0720-56-2246)
Motoko Naruse Motoko Naruse Design Office, 2-7-6 Minami-Aoyama Minato-ku, Tokyo 107, Japan (03-401-4481)
Masahisa Nakamura 4-1-14-1307 Hiroo Shibuya-ku, Tokyo 150, Japan (03-797-6137)
Nice & Company 1F Daiken Bldg., 4-9-15 Wakabayashi Setagaya-ku, Tokyo 154, (03-487-0045)
Nice Creation Co., Ltd. 2-7-6 Oyodo-naka Oyodo-ku, Osaka 531, Japan (06-453-2851)
Masaharu Noguchi 1-21-1-302 Jingumae Shibuya-ku, Tokyo 150, Japan (03-405-6814)
Takaaki Nomura 4-1-18 Asahi-machi Maebashi City, Gunma 371, Japan (0272-43-7061)
North Marine Drive Co., Ltd. 103 Minamiaoyama 131-Bldg., 7-4-17 Minami-Aoyama Minato-ku, Tokyo 107, Japan (03-406-8045)
NYLON 3-3-36 Miyahara Yodogawa-ku, Osaka 532, Japan (06-391-4951)
Kohzo Okada 2-A Fortona, 7-12-23 Roppongi Minato-ku, Tokyo 106, Japan (03-408-8664)
Akio Okumura Packaging Create Inc., 2F Daito Bldg., 5-13-11 Nishi-Tenma Kita-ku, Osaka 530, Japan (06-312-7978)
Yosuke Onishi 3B Yuki Flat, 1-14-15 Nishi-Azabu Minato-ku, Tokyo 106, Japan (03-408-0841)
Ota Yuji Sun-ad Co., Ltd., 9F Palace Bldg., 1-1-1 Marunouchi Chiyoda-ku, Tokyo 100, Japan (03-274-5021)
Ozaki Paper Products Co., Ltd. 1-15-15 Kuwatsu Sumiyoshi-ku, Osaka 558, Japan (06-719-2967)
Ozone Community Co., Ltd. 3-12-6 Sendagaya Shibuya-ku, Tokyo 151, Japan (03-478-8471)
Paper Moon Int'l Japan, Inc. 2-31-7-802 Jingumae Shibuya-ku, Tokyo 150, Japan (03-478-4320)
Papyrus Company Ltd. 1F Satomi Bldg., 2-24-4 Jingumae Shibuya-ku, Tokyo 150, Japan (03-402-5401)
Peanut Boy 4-26-24 Jingumae Shibuya-ku, Tokyo 150, Japan (03-479-0780)
Pink Bus 13 Udagawa-cho Shibuya-ku, Tokyo 150, Japan (03-770-2070)
Planet, Inc. 2-16-14 Yanagibashi Taito-ku, Tokyo 111, Japan (03-865-0991)
Pop Shop Enterprise Co., Ltd. 3-11-12 Minami-Aoyama Minato-ku, Tokyo 107, Japan (03-408-0207)
Portfino 1-1-3 Shibata Kita-ku, Osaka 530, Japan (06-374-0009)
Pound House 29 Nakanocho Daihoji-cho Minami-ku, Osaka 542, Japan (06-251-4334)
Printemps Ginza S.A. 3-2-1 Ginza Chuo-ku, Tokyo 104, Japan (03-567-0077)
Pro-Line Inc. NTC Bldg., 3-21-17 Akasaka Minato-ku, Tokyo 107, Japan (03-587-0271)
Raika Co., Ltd. 2-8-17 Kita-dori Miyakojima Miyakojima-ku, Osaka 534, Japan (06-922-8423)
Richard Co., Ltd. 2-31 Shinsaibashi-suji Minami-ku, Osaka 542, Japan (06-211-4477)
Rock-za Co., Ltd. 2-1 Kanda-Ogawamachi Chiyoda-ku, Tokyo 101, Japan (03-295-1545)
Kichiro Saito Hi-Production Co., Ltd., 205 Mansion Saien, 3-43 Kaiunbashi-dori Morioka City, Iwate 020, Japan (0196-52-1519)
Katsuichi Sakakibara Design Studio 202, 405 Asahi-Rokubancho Mansion, 4 Rokubancho Chiyoda-ku, Tokyo 102, Japan (03-221-9481)
Eiichi Sakota 20-5 Hirai-sanso Takarazuka City, Hyogo 665, Japan (0797-89-2257)
Sanrio Company, Ltd. 1-6-1 Osaki Shinagawa-ku, Tokyo 141, Japan (03-779-8111)
Sanyo Electric Co., Ltd. 100 Dainichihigashi-cho Moriguchi City, Osaka 570, Japan (06-901-1111)
Masakazu Sawa Maguna, Inc., 6-13-3 Ginza Chuo-ku, Tokyo 104, Japan (03-543-4341)
Helmut Schmid 3-24-14-707 Tarumi-cho Suita City, Osaka 564, Japan (06-338-5566)
Takahiro Shima Shima Design Office Inc., 8F Daido-seimei Bldg., 1-22 Edobori Nishi-ku, Osaka 550, Japan (06-441-8188)

Shimojima Shoji Co., Ltd. 5-29-8 Asakusabashi Taito-ku, Tokyo 111, Japan (03-861-6059)
Shin & Company France-ya Bldg., 5-13-1 Nishi-Gotanda Shinagawa-ku, Tokyo 141, Japan (03-495-6691)
Shinjuku Takano Co., Ltd. 3-26-11 Shinjuku, Shinjuku-ku, Tokyo 160, Japan (03-354-0222)
Kokichi Shirakawa Kokichi Shirakawa Design Room, 300 Green Heights, 142 Showa-machi Takasaki City, Gunma 370, Japan (0273-62-1318)
Shiseido Boutique The Ginza Co., Ltd. 5F Gyosei Bldg., 7-4-12 Ginza Chuo-ku, Tokyo 104, Japan (03-572-2121)
Sony Creative Products Inc. 3-6 Kinoi-cho Chiyoda-ku, Tokyo 102, Japan (03-237-5645)
Sony Plaza Co., Ltd. Komatsu Annex Bldg., 6-8-5 Ginza Chuo-ku, Tokyo 104, Japan (03-575-2511)
Suger Creative Production 303 Daikanyama Pension, 2-16-13 Ebisu-nishi Shibuya-ku, Tokyo 150, Japan (03-461-8367)
Sun Design Associates 7F Taisei Bldg., 1-29 Minami-Kyutaro-cho Higashi-ku, Osaka 541, Japan (06-261-2961)
Sunshine City Corporation 3-1 Higashi-Ikebukuro Toshima-ku, Tokyo 170, Japan (03-989-3531)
Suntory Limited 2-1-40 Dojimahama Kita-ku, Osaka 530, Japan (06-346-1140)
Surplus Inc. 306 Green Town, 5-28-7 Jingumae Shibuya-ku, Tokyo 150, Japan (03-400-5049)
Masaru Suzuki 2-3-105 Tsujimachi-jutaku, 1-8 Tsujimachi Kita-ku, Nagoya 462, Japan (052-915-1387)
Zenpaku Suzuki B·BI Studio Inc., 3F Akiba Bldg., 3-12-14 Ginza Chuo-ku, Tokyo 104, Japan (03-546-0918)
Nario Taiten T-Box, 5-211 Katsuragawa Heights, 1-1 Ishihara-Osada-cho Kishoin Minami-ku, Kyoto 601, Japan (075-661-2960)
Yukichi Takada I.F.Planning Co., Ltd., 1-33 Kitahama Higashi-ku, Osaka 541, Japan (06-231-8282)
Takashimaya Co., Ltd. 2-4-1 Nihonbashi Chuo-ku, Tokyo 103, Japan (03-211-4111)
Jun Takechi Verve Inc., 1-26-12-1001 Shinjuku, Shinjuku-ku, Tokyo 160, Japan (03-341-3649)
Akira Takeshita Takeshita Creative Director Co., Ltd., 7-6-41-308 Akasaka Minato-ku, Tokyo 107, Japan (03-589-0913)
Jiro Takeuchi Micky House Inc., North-4 West-15-chome Chuo-ku, Sapporo 060, Japan (011-644-2543)
Yasuo Tanaka Pakage Land Co., Ltd., 201 Tezukayama Tower Plaza, 1-3-2 Tezukayamanaka Sumiyoshi-ku, Osaka 558, Japan (06-675-0138)
Osamu Tatematsu Decom Creative Art & Service Co., Ltd., 6F Lion Bldg., 3-19-1 Marunouchi Naka-ku, Nagoya 460, Japan (052-951-6591)
TDK Corporation 1-13-1 Nihon-bashi Chuo-ku, Tokyo 103, Japan (03-278-5043)
The Design Associates Maruyama Bldg., 2-3-8 Azabu-dai Minato-ku, Tokyo 106, Japan (03-587-1740)
Titicaca 6-20-104 Udagawa-cho Shibuya-ku, Tokyo 150, Japan (03-476-3449)
Masatoshi Toda Toda Studio, 307 Crown Tsukiji, 7-16-3 Tsukiji Chuo-ku, Tokyo 104, Japan (03-545-1533)
Taiki Toriyama Bird Design House, 6-6-205 Yoriki-cho Kita-ku, Osaka 530, Japan (06-354-1797)
Try Co., Ltd. Harajuku Aoi Mansion, 3-4-10 Sendagaya Shibuya-ku, Tokyo 151, Japan (03-401-4337)
Kayaki Tsukamoto Communication Arts R, 5F Rokuwa Bldg., 5-1-4 Roppongi Minato-ku, Tokyo 106, Japan (03-405-0270)
Uchu-Hyakka 1F Kuretake Bldg., 4-9 Udagawa-cho Shibuya-ku, Tokyo 150, Japan (03-464-4148)
Masaru Uda U Design Office, 1-4-22-602 Moto-machi Naniwa-ku, Osaka 556, Japan (06-647-3711)
USPP (United States of Paradise Park) 3-41-14 Sendagaya Shibuya-ku, Tokyo 151, Japan (03-478-4120)
Vernal and Company 707 Royal Ichibancho, 6-1 Ichiban-cho Chiyoda-ku, Tokyo 102, Japan (03-234-7533)
Voxel Yamamoto 55 Daikoku-cho Kawaramachi-Sanjo Nakagyo-ku, Kyoto 604, Japan (075-221-0969)
Wachi Field Co., Ltd. 2-19-5 Jiyugaoka Meguro-ku, Tokyo 152, Japan (03-725-0881)
Wakana Co., Ltd. 43F Sunshine 60, 3-1-1 Higashi-Ikebukuro Toshima-ku, Tokyo 170, Japan (03-989-5962)
Washington Shoe Co., Ltd. 5-7-7 Ginza Chuo-ku, Tokyo 104, Japan (03-572-5911)
Weekends 85 Atrium, 1-8-21 Jingumae Shibuya-ku, Tokyo 150, Japan (03-401-8350)
Wonder House 11-2 Udagawa-cho Shibuya-ku, Tokyo 150, Japan (03-461-6266)
Wooden Doll 1-6-6 Jingumae Shibuya-ku, Tokyo 150, Japan (03-423-0417)
World Co., Ltd. 6-8-1 Minatojima-Nakamachi Chuo-ku Kobe City, Hyogo 650, Japan (078-302-3111)
Takeo Yagi Office P&C Inc., A 406 Villa Modelna, 1-3-18 Shibuya Shibuya-ku, Tokyo 150, Japan (03-406-0191)
Kijiro Yahagi 1-39-1-203 Hon-cho Shibuya-ku, Tokyo 151, Japan (03-375-9204)
Yamamoto Kansai Co., Ltd. Tokyo Central Omotesando Bldg., 4-3-15 Jingumae Shibuya-ku, Tokyo 150, Japan (03-470-2030)
Masanori Yamamoto Design Center, Bridgstone Corpration, 5F AXIS Bldg., 5-17-1 Roppongi Minato-ku, Tokyo 106, Japan (03-583-0311)
Shun'yo Yamauchi Shun'yo Yamauchi Design Office, 305 Yotsuya Bldg., 3-26 Yotsuya-dori Chigusa-ku, Nagoya 464, Japan (052-782-6989)
Tatsuo Yamazaki Advertising Section, Maruzen Co., Ltd., Dai-ni Maruzen Building, 3-9-2 Nihonbashi Chuo-ku, Tokyo 103, Japan (03-272-7033)
Takeo Yao YAO Design Institute Inc., 3F Shihoshoshikaikan, 9-3 Honshio-cho Shinjuku-ku, Tokyo 160, Japan (03-357-3686)
Kazuo Yasuhara Advertising Dept., Shiseido Co., Ltd., 7-5-5 Ginza Chuo-ku, Tokyo 104, Japan (03-572-5111)
Y.Inohana & Sons Shinkyogoku-Sanjo Nakagyo-ku, Kyoto 604, Japan (075-221-3561)
Yokohama Okadaya Co., Ltd.(Okadaya More's) 1-3-1 Minami-Saiwai Nishi-ku, Yokohama 220, Japan (045-311-1471)
Zero One Co., Ltd. 1F Hara Bldg., 6-10-10 Jingumae Shibuya-ku, Tokyo 150, Japan (03-486-3601)
Zoobon 1F Mizutani Bldg., 35-9 Higashi-Shimizu-cho Minami-ku, Osaka 542, Japan (06-241-1808)
[Canada]
Rolf Harder Rolf Harder & Assoc. Inc., 1350 Sherbrooke St. W., #1000 Montréal, Québec H3G 1J1, Canada (514-281-1093)
Burton Kramer Burton Kramer Associates Ltd., 20 Prince Arthur Avenue, Suite 1E Toronto, Ontario M5R LB1, Canada (416-921-1078)
Fernando Medina Medina Design, 3001 Sherbrooke West, Suite 1105 Montréal, Québec H3Z 2X8, Canada (514-935-8092)
[United States]
R.O.Blechman R.O.Blechman Inc., 2 West 47th St. New York, NY 10036, USA (212-869-1630)
Chuck Byrne Chuck Byrne Design, 5528 Lawton Ave. Oakland, CA 94618, USA (415-658-6996)
Chermayeff & Geismar Associates 15 East 26th St. New York, NY 10010, USA (212-532-4499)
Robert P. Gersin Robert P. Gersin Associates, Inc., 11 East 22 Street New York, NY 10010, USA (212-777-9500)
Engene Grossman Anspach Grossman Portugal Inc., 711 Third Ave. New York, NY 10017, USA (212-692-9000)
Peter Harrison Pentagram Design Ltd., 212-5th Avenue 17th Floor New York, NY 10010, USA (212-683-7000)
Malcolm Grear Designers, Inc. 391 Eddy Street Providence, RI 02903, USA (401-331-2891)
Michael Manwaring The Office of Michael Manwaring, 1045 Sansome St. San Francisco, CA 94111, USA (415-421-3595)
Michael Mabry Design 212 Sutter Street San Francisco, CA 94108, USA (415-982-7336)
Nordstrom Arcade Plaza Bldg., 1321 2nd Ave., Seattle, WA 98101, USA (206-343-6814)
Woody Pirtle Pirtle Design, 4528 Mckinney #104 Dallas, Texas 75205, USA (214-522-7520)
Jack Summerford Summerford Design, Inc., 2706 Fairmount Dallas, Texas 75201, USA (214-748-4638)
Tim Girvin Design, Inc. 911 Western #408 Seattle, WA 98104, USA (206-623-7808)

Fred Troller Fred Troller Associates, 12 Harbor Lane Rye, NY 10580, USA (914-698-1405)
Geoge Tscherny Geoge Tscherny, Inc., 238 East 72 Street New York, NY 10021, USA (212-734-3277)
Elbert K. Tsuchimoto K & E Graphics, Inc., 1179 Palekaiko St. Pearl City, Hawaii 96782, USA
Ryo Urano UCI Inc., 1088 Bishop Street, Suite 1226 Honolulu, Hawaii 96813, USA (808-533-4296)
Michael Vanderbyl Vanderbyl Design, 539 Bryant Street 4th Fl., San Francisco, CA 94107, USA (415-543-8447)
Massimo Vignelli Vignelli Associates, 475 Tenth Avenue New York, NY 10018, USA (212-244-1919)
Tamotsu Yagi Esprit de Corporation, 900 Minnesota St. San Francisco, CA 94107, USA (415-648-6900)

[Australia]

Raymond Bennett Raymond Bennett Design Assoc., 9 Myrtle St., Crows Nest, NSW 2605, Australia (02-959-5777)
Ken Cato Cato Design Inc. Pty. Limited, 254 Swan St. Richmond, Victoria 3121, Australia (03-429-6577)
Mimmo Cozzolino Cozzolino/Ellett, PO Box 198 Heidelberg, Victoria 3084, Australia (03-458-1638)
Keith Davis Davis Farrell+Associates, 285 Clarence St. Sydney, NSW 2000, Australia (02-264-7466)
Garry Emery Emery Vincent Associates, 80 Market St., South Melborne, Victoria 3205, Australia (03-699-3822)
Barrie Tucker Barrie Tucker Design Pty. Ltd., 114 Rundle Street, Kent Town Adelaide, Sth Aust. 5067, Australia (08-42-7834)

Hong Kong

Alan Chan Alan Chan Design Co., 801/802, 8/F., Shiu Lam Building, 23 Luard Road, Wanchai, Hong Kong (5-278228)
Henry Steiner Graphic Communication Ltd., 28c Conduit Road, Hong Kong (5-485548)
Kan Tai-keung SS Design & Production, 22/F., 276-278 Lockhart Road, Wanchai, Hong Kong (5-748399)

[India]

Benoy Sarkar HHEC Design Cell, 11 A Rouse Avenue Lane, New Delhi 110002, India (3311086)

[Israel]

Dan Reisinger Studio Reisinger, 5 Zlocisti Street Tel-Aviv 62994, Israel (03-251433, 03-266723)

[England]

Mervyn Kurlansky Pentagram Design Ltd., ll Needham Road London W11 2RP, England (01-229-3477)
Minale Tattersfield+Partners Ltd. Burston House, Burston Road, Putney London SW15 6AR, England (01-788-8261)
David Pocknell Pocknell & Co., Readings Farm, Blackmore End Braintree, Essex CM7 4DH, England (0787-61207)
Trikett & Webb Ltd. 84 Marchmont Street London W1N 1HE, England (01-388-5832)

[The Netherlands]

Anthon Beeke Anthon Beeke & Associates, Hortusplantsoen 6, 1015 T2 Amsterdam, The Netherlands (0-20-220600)
Marianne Vos Samenwerkende Ontwerpers, Herengracht 160, 1016 BN Amsterdam, The Netherlands (020-240547)

[Switzerland]

Karl Domenic Geissbühler Theaterstrasse 10 Zürich 8001, Switzerland (01-476761)
Fritz Gottschalk Gottschalk+Ash Int'l, Sonnhaldenstrasse 3, 8032 Zürich, Switzerland (01-252-5042)
Wirth Kurt Bürglenstrasse 21 Bern CH, Switzerland (44 63 88)
Armin Vogt Armin Vogt Partner, Münsterplatz 8, Basel, Switzerland (061-258385)

[Spain]

America Sanchez Aribau 245-E3, 08021 Barcelona, Spain (201-8867)

[Italy]

Leonardo Baglioni Studio Leonardo Baglioni, Via Della Torre Del Gallo, 26 Firenze 50125, Italy (055-229192)
Walter Ballmer Unidesign, Via Revere 16 Milano 1-20123, Italy (02-4694769)
Giorgio Gregori Alchimia, Via Fratelli Gabba 5, 20131 Milano, Italy (02-8052476)
Heinz Waibl Signo SRL, Via Emanuele Filiberto 14, 20149 Milano, Italy (02-3313352, 02-3313347)

[Austria]

Harry Metzler Harry Metzler Artdesign, Brand 774, A-6867 Schwarzenberg, Austria (0 55 12 /34 94)

[West Germany]

Pierre Mendell Mendell & Oberer, Widenmayerstr. 12, 8000 München 22, W.Germany (089-224055)
Anton Stankowski Stankowski+Duschek, Lenbachstrasse 43 D-7000 Stuttgart 30, W.Germany (0711-814408)

[Denmark]

Flemming Nielsen Nielsen & Baillie, Ny Adelgade 5A, DK 1104 Copenhagen, Denmark (45 1 15 96 02)

[Norway]

Bruno Oldani Bygdøy allé 28B 0265 Oslo 2, Norway (02-390126)

[Yugoslavia]

Edi Berk Studio KROG, Krakovski nasip 22 Ljubljana 61000, Yugoslavia (38-61-210 051)
Boris Ljubičić Studio International, Buconjićeva 43/II Zagreb 41000, Yugoslavia (003841572706)

索引
社名/店名

Index
companies/stores/shops

A
ア・ストア・ロボット　A Store Robot Co.,Ltd.　87, 436
アクロス　Across Co., Ltd.　241
赤坂青野　Akasaka-Aono　314
赤坂米穀　Akasaka Beikoku Co., Ltd　76
明石企画　Akasi Kikaku Co., Ltd.　108
Alicia Co., Ltd.　154
Ami　241
アンドア　Andor　165
アンジェラ・グラショフ　Angela Grashoff　79
アンガス・アンド・クート　Angus & Coote Pty. Ltd.　345
アン　ann　344
アナザーワン　Another One Co., Ltd.　273
アオイ　Aoi Co., Ltd.
フェンディ　Fendy　334
サルバドーレ・フェラガモ　Salvadore Ferragamo　453
青山商品研究所　Aoyama Research Institute of Fashion Marchandising Co., Ltd.　361
アローム　Arome Co., Ltd.　252
アジアの靴　Asian Shoes, Co., Ltd.　106
ASNN　172
アトリエ伊藤佐智子　Atelier Ito Sachiko　163
アトリエ・ニキティキ　Atelier Niki Tiki Co., Ltd.　276
ATTIC　207
オーストラリア郵政省　Australia Post　159
B
ベーリー・ストックマン　Bailey Stockman Co., Ltd.　26, 132, 138
バルーン　Ballon Co., Ltd.　240
バルツァース・アクティエンゲゼルシャフト　Balzers Aktiengesellschaft　169
バナナレコード　Banana Record　379
バーバラ・ペテルカ　Barbara Peterca, Hand Weaving　373
バーニーズ・ニューヨーク　Barney's New York　110
バッサーニ・ティチーノ　Bassani Ticino　198, 310
B-By-Three Co., Ltd.　400
ビームス　Beams Co., Ltd.　367, 368
Beat Pops By Jemmy's　278
別海漁業協同組合　Bekkai Fishery Cooperative Assoc.　327
ベル(神戸ベル)　Belle Co., Ltd.(Kobe Belle)　330, 331
バークレイ　Berkeley Co., Ltd.　136
ビックカメラ　Bic Camera Co., Ltd.　166
ビッキー　Bikky　70
ビアンクール　Billancourt Limited　113
ビザール　Bizarre　245
ブルーミングデイルズ　Bloomingdale's　49, 50, 52, 73, 74, 83, 129, 246, 332
ブルー＆ホワイト　Blue & White　18
ブルーグラス(ボザール)　Blue Glass (Beaux-arts)　32
ザ・ブーツ・カンパニー　The Boots Company　331
ブティック・バザール　Boutique Bazaar　235
ブティック・ギルズ　Boutique Gilles　380
ブリヂストン　Bridgestone Corporation　450
ブッチ　Bucci　313
ブッホハンドルンク・シェルツ　Buchhandlung Scherz　191
ブルクハルト・ラジオ・テレビジョン　Burkhart AG Radio-Television　170
C
Can Co., Ltd.　257, 277
カレン・チャールズ　Caren Charles　243
カルロス・ファルチ・ジャパン　Carlos Falchi Japan Co., Ltd.　204
キャスネル　Casnel Ltd.　363
チャンプ・カンパニー　Champ Company　392
チェース・マンハッタン・バンク　Chase Manhattan Bank　3
シェトワ　Chez Toi Co., Ltd.　78
チハラ　Chihara Co., Ltd.　187
Chix　451
ショコラティエ・マサール　Chocolatier Masále　133
シンシナティ現代美術センター　Cincinnati Contemporary Arts Center　196
CMらんど　CM Land Co., Ltd.　362
クラブ・ジャップス　Club Japs Inc.　121
コサージュ　Corsage　312
クロアチア　Croatia　353
シンシア　Cynthia Corp. Ltd.　146
D
大中　The Dai'chu, Inc.　441, 442
大日本スクリーン製造　Dainippon Screen Manufacturing Co., Ltd.　296
大和シュガル　→ Sugal Creative Production
ダラス・マーケット・センター(インフォワークス)　Dallas Market Center (Infoworks)　46
ダーリン・ハーバー・オーソリティ　Darling Harbour Authority　80
デタント　Detente　402
ディボア　Diboa　260
ダイバ・ソフト・ファーニシングス　Diva Soft Furnishings　376
デライツ・クラブ　D'Lites Club
リトル・スリーク・ビーン　Little Sleek Bean　219, 430
オクトパス・アーミー　Octpus Army　210
イエロー・モンスター　Yellow Monster　220
ドノバン(モノショップ・フィール)　Donoban Inc. (Mono-shop "Feel")　303
Dourakuya Grp. (D. Shop Co., Ltd.)　17
ダスティー・ミラー　Dusty Miller Co., Ltd.　274
E
エゴ　Ego Hair Styling And Beauty Therapy　346
ジ・エンポリアム　The Emporium　385
エスプリ・デ・コーポレーション　Esprit De Corporation　125, 126, 127, 128, 287
F
FIF Group　269
フランドル　Flandre Co., Ltd.　104
フォーライフ・レコード　For Life Record Co., Ltd.　405
45 RPM-Studio Co.　151
フラテッリ・ロセッティ　Fratelli Rossetti　178, 179
Free Hand　7
フリーウェイ　Freeway Co., Ltd.　387
フロンティア　Frontier　119
フジッコ　Fujikko Co., Ltd.　352
藤丸　Fujimaru Co., Ltd.　404
フレスト　Furest　423
ふるさとや　Furusatoya　58
G
ジーピー　G &P Co., Ltd.
GP-8 渋谷店　GP-8 Shibuya　105
パジャブ渋谷店　Pajaboo Shibuya 149
ガスライト・レコード・アンド・テープ　Gaslight Records & Tapes　366
ガスト・ハウス　Gast Haus　391
現代衣学　Gendai-Igaku Co., Ltd.　275
Gene Kelly　429
ジェラルド・D・ハイネス・インタレスツ　Gerald D. Hines Interests　28
ジャイアント・ストアーズ・リミテッド　Giant Stores Limited　374, 375
銀座小松　Ginza Komatsu　176
銀座千疋屋　Ginza Senbikiya　36, 302
グラマー　Glamour　389
Gordini　452
グルモン　Gourmand　428
グレース・ブラザーズ・パーティ・リミテッド　Grace Bro. Pty. Ltd. 297,365
グランド・キャニオン　Grand Canyon Co., Ltd. 212
グラフィーシェ・ザムルンク・アルバーティナ　Graphische Sammlung Albertina　139
グリーンハウス　Green House Co., Ltd. 253
グリーン・キング・アンド・サン　Greene King & Son PIC　382
グレート・ブリテン　Great Britain Corporation　85, 429
グレタ・ガルボ　Greta Galubot　93, 94
ジーセンコンフェクト　G-sen Confect Co., Ltd.　299
H
インド手織物・手工芸品輸出会社　Handlooms & Handicrafts Export Corp. of India Ltd.　202
阪神百貨店　Hanshin Department Store Co., Ltd.　410
阪神商事　Hanshinshoji Co., Ltd.　165
博品館トイ・パーク　Hakuhinkan Toy Park　393
原宿サーキット　Harajuku Circuit　115
原宿ル・ポンテ　Harajuku Le Ponte　8
ハロッズ・ナイトブリッジ・リミテッド　Harrods Knightbridge Ltd.

244, 383
ハリーズ　Harry's　157
ハタダ　Hatada Co., Ltd.　290
林原グループ　Hayasibara Group　111
ヘルス・フード・ストアー　Health Food Store Ltd.　194
広島民芸社　Hiroshimamingei-sha Co., Ltd.　316
ホビーショップ・サンドゥ　Hobby Shop Sando　455
ホームズ・アンダーウェア　Homes' Underwears　153
ハニー　Honey Co., Ltd.
カドリー・ブラウン　Cuddly Brown　371
ハラッパA　Harrapa A　107, 343
キネティクス　Kinetics　98
T.P.O.　97
ホーユー　Hoyu Co., Ltd.　439
I
日本IBM　IBM Japan　283
アイドル　Idol Co., Ltd.　395
アイドル・メイクレット　Idol Makelet Co., Ltd.　337
池袋ショッピングパーク　Ikebukuro Shopping Park Co., Ltd. 183
イマジネイリアム　Imaginarium　236
イマーゴ　Imago　51
INAX Corporation　65
イング・セブン　Ing Seven Co., Ltd.　416
井上洋紙店　Inoue Yoshiten Co., Ltd.　237
インタークルー　Intercrew　190
インターナショナル・デザイン・センター/ニューヨーク International Design Center/New York　142
イルマ・スーパーマーケット　Irma Supermarket　424
伊勢丹　Isetan Department Store Co., Ltd.　357, 443, 445
イッセイミヤケ・アンド・アソシエーツ　Issey Miyake & Associates Inc.　99
イッセイミヤケインターナショナル　Issey Miyake International Inc.　2, 22, 23, 24, 186
岩田屋　Iwataya Department Store Co., Ltd.　131, 268
J
日本舟艇工業会　Japan Boating Industry Assoc.　226
宝石の富士屋　Jewelry Fujiya　377
ジョイント　Joint Co., Ltd.　101
ジョイス・ブティック　Joyce Boutique Ltd.　25
ジュン　Jun Co., Ltd.　88, 418, 447
Junko Shimada Part 2　39
ジュロン・バード・パーク　Jurong Bird Park　38
K
花外楼　Kagairoh Co., Ltd.　292
柿安本店　Kakiyasu-Honten　325
上宝村役場(岐阜県吉城郡)　Kamitakara-mura Village Office　425
鹿六　Karoku Co., Ltd.　372
京阪百貨店　Keihan Department Stores Ltd.　116, 118
京王百貨店　Keio Department Store Co., Ltd.　298
ケンウッド　Kenwood Corporation　355
Kenzo Paris　349
キデイランド　Kiddy Land Co., Ltd.　364
Kid's Monster Inc.　258
キジュウロウ・ヤハギ　Kijuro Yahagi Co., Ltd.　43
キムラタン　Kimuratan Co., Ltd.　203
近鉄百貨店　Kintetsu Department Store Co., Ltd.　61
キタイ　Kitaj　200
コンガ　Konga　234
コーセー化粧品　Kose Cosmetics Co., Ltd.　158, 180, 182
寿精版印刷　Kotobuki Seihan Printing Co., Ltd.　304
K. Shop, Takako Kubo　336
L
ラフェーテ　Laféte Co., Ltd.　291
ラフォーレ原宿　Laforet Harajuku　13, 14
Lamp Fox　120
ラパス　Lapaz　398
ラリオ　Lario　440
ラリー・ジュエリー　Larry Jewelry　294
ルシアン・プランニング　Lecien Planning Corporation　39
リーザー　Leeser　269
リトルアンデルセン　Little Andersen Co., Ltd.
ヒステリック・ミニ　Hysteric Mini　103
チャビー・ギャング　Chaby Gang　370
ロス　Los　147
ランバージャック　Lumber Jack　454
リリック　Lyric Co., Ltd.　438
M
毎日新聞社　Mainichi Shimbunsha　224
マークトガッセ・デュベンドルフ　Marktgasse Dübendolf　161
丸善　Maruzen Co., Ltd.　175
マトリキュラ　Matricula　63
松城紙袋工業　Matsushiro Papar Bag Industry Co., Ltd. 62, 89, 249
松屋　Matsuya Co., Ltd. 20
松坂屋　Matsuzakaya Co., Ltd.　173, 228, 285, 286, 300, 432
ザ・メイ・デパートメント・ストアー　The May Department Store Co.　69
メディア・ショップ　Media Shop　372
目黒区美術館　Meguro Museum of Art' Tokyo　358
メンズショップ・マルヤマ　Men's Shop Maruyama　350
メンズショップ・トラヤ　Men's Shop Toraya　434
メンズショップ・トラヤ(Lamp Fox)　Men's Shop Toraya (Lamp Fox)　120
メッセ・フランクフルト　Messe Frankfurt GMBH　64
ミドリヤ　Midoriya Co., Ltd.　411
三越　Mitsukoshi Ltd.　9, 91, 261, 263, 264
光村原色版印刷所　Mitsumura Printing Co., Ltd.　181
ミクサージュ　Mix Age　203
宮坂醸造　Miyasaka Jyozo Co., Ltd.　317
モン・ジャルダン　Mon Jardin　444
モーニン・ムー　Mornin'Mou　342
モロゾフ　Morozoff Limited　189, 259
元町ポーター・クラブ　Motomachi Porter Club　117
ムパタ　Mpata Shop　212
ニューヨーク近代美術館　The Museum of Modern Art, New York　27, 37, 282
N
ナショナル・カウンシル・オブ・ステート・ガーデン・クラブ National Council State Garden Clubs Inc.　422
ナショナル・ウェストミンスター・バンク　National Westminster Bank　66, 67, 68
ねぼけ堂　Nebokedo　293
NECアベニュー　NEC Avenue Co., Ltd.　397
ニーマン・マーカス　Neiman Marcus　281
Nice　351
ナイス・クリエーション　Nice Creation Co., Ltd. 402
Nice & Company　54, 437
日本経済新聞社　Nihon Keizai Shimbun, Inc.　34, 90
ニッカウヰスキー　The Nikka Whisky Distilling, Co., Ltd. 160
日本通運(遠野営業所)　Nippon Express Co., Ltd.　320
日本レジャー開発　Nippon Leisure Kaihatsu Co., Ltd.　318
ノードストローム　Nordstrom　75, 82
ノース・マリン・ドライブ　North Marine Drive Co., Ltd.　206
NYLON　222
O
オークヴィル・グロッサリー　Oakville Grocery Co.　339
小田急百貨店　Odakyu Department Stores Co., Ltd.　192
オオコシ　Ohkoshi Co., Ltd.　60
岡田屋モアーズ　Okadaya-More's　10, 11, 12, 29, 100, 211
大宮ステーションビル　Omiya Station Building　40, 141, 230
オレンジ・ビレッジ・クラブ　Orange Village Club Co., Ltd. 409
オゾン・コミュニティー　Ozone Community Co., Ltd.　150
P
ペイス　Pace　156
ペーパー・ムーン・インターナショナル・ジャパン　Paper Moon Int'l Japan, Inc.　86
パピルス・カンパニー　Papyrus Co., Ltd.　417
パルコ　Parco, Inc.　137
ピーナッツ・ボーイ　Peanut Boy　255
ペパーミント　Peppermint Co., Ltd.　278
ペパーミント・パラレル　Peppermint Parallel　16
ピアージュ　Piage　30
パイル　Pile Co., Ltd.　449
ピンクバス　Pink Bus　381
ポップ・インターナショナル　Pop International Co., Ltd.　341